A Skeptic's Guide to the Mind

Also by Robert A. Burton, M.D.

On Being Certain: Believing You Are Right Even When You're Not
Doc-in-a-Box
Cellmates

A Skeptic's Guide to the Mind

**What Neuroscience
Can and Cannot
Tell Us About Ourselves**

ROBERT A. BURTON, M.D.

ST. MARTIN'S PRESS
NEW YORK

To Adrianne

www.stmartins.com

Designed by Maura Rosenthal/MSpace

Library of Congress Cataloging-in-Publication Data

Burton, Robert Alan, 1941–
    A skeptic's guide to the mind : what neuroscience
can and cannot tell us about ourselves / Robert A.
Burton, M.D.
        p. cm.
    Includes bibliographical references and index.
    ISBN 978-1-250-00185-6 (hardcover)
    ISBN 978-1-250-02840-2 (e-book)
  1. Neurosciences.   2. Brain—Physiology.
3. Mind and body.   I. Title.
    RC343.B87 2013
    616.8—dc23

                                    2012041265

First Edition: April 2013

10 9 8 7 6 5 4 3 2 1

Contents

A Skeptic's Guide to the Mind

Introduction

A book is the only place in which you can examine a
fragile thought without breaking it, or explore an explo-
sive idea without fear it will go off in your face. It is one
of the few havens remaining where a man's mind can
get both provocation and privacy.

—attributed to Edward P. Morgan[1]

E ach of us has a pretty good sense of what a mind is. It's
that indescribably vague, palpable, yet invisible "something" just behind the forehead that is responsible for our
thoughts. Beyond that, all bets are off. Some say it is simply the
software for the brain, or what the brain does. Others have a
more cosmic view of a mind without boundaries, or a nonmaterial essence that transcends and survives the death of the
body. For most of us, it is both the measure of the man and the
tool whereby we make this measurement. In turn the value of
this judgment depends on how we believe that our mind
works—how much of our thought and behavior is dictated by
underlying biological predispositions and involuntary unconscious brain activity and how much is within our conscious
control.

The consequences of this determination are enormous at
both a personal and a global level. From attributing motivation and assigning personal responsibility, to assessing the
threat of a nuclear attack by North Korea or Iran, we are

constantly being asked to read our minds and the minds of others. And yet, we have no idea what a mind actually is. Despite 2,500 years of contemplation and the more recent phenomenal advances in basic neuroscience, the gap between what the brain does and what the mind experiences remains uncharted territory. Though many scientists would like to believe that this gap can be fully bridged with further scientific advances, they are mistaken.

Science is the only method we have for establishing the factual basis of what the mind might be. But how do you adequately investigate something that can't be measured? Understanding how the brain works is great for describing biological functions, but still leaves us guessing as to what is being consciously experienced. Looking at the most detailed brain scans won't capture what we feel when we experience love or despair any more than examining the individual pixels in a Chuck Close painting will give you an overall sense of the painting. (To underscore how little we really know about the mind, we need only realize that some prominent philosophers still seriously debate whether or not the mind even plays a role in our behavior.)

Nevertheless, with powerful new imaging techniques such as the functional magnetic resonance imaging (fMRI) scan, cognitive science has become the de facto mode of explanation of behavior, rushing into the vacuum created by the failure of previous psychological and philosophical theories to fulfill their initial promise. Neuroscience is now seen as the preeminent model of the mind and the creator and guardian of our cultural mythology. It has garnered the ultimate status of a becoming a prefix. A new language is emerging: neuro-economics, neuro-aesthetics, neuro-theology, neuro-innovation, neuro-linguistics, neuro-marketing, neuro-networking. Philosophers commonly cite neurological case studies as evidence for their theories. Market crashes are explained by fMRIs. Neuroscientists tell us why we prefer Coke to Pepsi.

Such advances have been seductive to the academic community and the general public. What once were privately acknowledged among neurologists as metaphysical musings are increasingly being offered and seen as scientifically based facts. Like a child handed a new toy, the scientific community isn't likely to proceed with caution.[2]

The race is on; the holy grail of science (and much of philosophy of mind) is to explain how a brain creates a mind. But the lack of rock-solid initial assumptions and consensus opinion as to what the mind "is" has resulted in a disjointed flurry of often unsupportable or contradictory behavioral observations. Try opening a newspaper or magazine without being confronted with yet another neuroscience tidbit being offered as an explanation of our behavior. Every day I see the most complex aspects of human behavior reduced to incomprehensible sound bites. For example, consider the number of dubious assumptions and logical inconsistencies necessary to come up with the recent headline in one of my favorite popular science magazines: "Possible Site of Free Will Found in Brain."[3] Or this British newspaper headline: "Bad Behavior Down to Genes, Not Poor Parenting."[4] While some observations are real advances, others are overreaching, unwarranted, wrongheaded, self-serving, or just plain ridiculous.

If this were merely of academic concern, I wouldn't bother with this book. What alarms me is that a lack of clear understanding of what we can and cannot say about the mind and the commonly held belief in the unlimited powers of science are a potent recipe for potential catastrophe. Those old enough will recall when psychoanalysis was touted as hard science and schizophrenia was attributed to an overbearing mother (the so-called schizophrenogenic mother). How about the immense suffering caused by those psychologists who un-critically conjured up the recovered memory syndrome with-out having a clear understanding of how memory works? Or the Nobel laureate António Egas Moniz's advancement of frontal lobotomies because patients were easier to manage? En-tire families were devastated by ideas that, at the time, seemed to make sense. Only in hindsight was the folly apparent.

And yet, like moths drawn to the flame, or amnesiacs who forget the lessons of history, brain scientists are repeating these same mistakes. Though it is easy to explain away their often unwarranted claims about the mind as arising from arrogance, greed, ignorance, or other psychological "quirks," this book will pursue a more basic premise: our brains possess involun-tary mechanisms that make unbiased thought impossible yet create the illusion that we are rational creatures capable of fully understanding the mind created by these same mechanisms.

Our brains have evolved piecemeal; contradiction, incon-sistency, and paradox are built into our cognitive machinery. We are hardwired to experience unjustified feelings about ourselves, our thoughts, and our actions; we possess an irre-pressible curiosity and desire to understand how the world works; we have developed an uncanny ability to see patterns whether or not they exist outside of our perceptions. Com-bine these traits with intrinsic cognitive constraints and you have the backdrop for modern-day neuroscience.

For me, the first step of any scientific inquiry should be a frank and open acknowledgment of the limitations of human thought, yet the recent spate of books and articles underscor-ing our inherent irrationality has done little to curb the excesses arising from a seemingly unshakable belief in pure reason. For neuroscientists and philosophers, as with the rest of us, the vis-ceral feeling of knowing you are right is far more convincing than the thought that we have limits to our powers of reason.

Such prominent neuroscientists as Antonio Damasio confidently proclaim that the explanation of consciousness is around the corner.[5] Philosopher Daniel Dennett has said, "I know of no reason to expect that a brain couldn't understand its own methods of functioning. Just because the brain is complex, with 100 million cells and a quadrillion synaptic connections, that doesn't mean we can't figure out what is going on within it."[6] Others like the late Nobel laureate Francis Crick are convinced that the brain and the mind are the same, and that we can use our minds to make this determination. Such grand predictions are likely to escalate as more enter the increasingly popular and competitive field of neuroscience. (In 2009, more than one thousand grad students received a Ph.D. in neuroscience, with many more Ph.D.'s granted in allied specialities such as psychology, adding to the tens of thousands already in the field.)

We are at a turning point in the history of self-understanding. Whether offering an fMRI scan as evidence of consciousness, claiming that we can use brain waves to read minds and detect lies, or positing that specific genes cause specific behaviors, neuroscience is in the process of redefining human nature. Science is trial and error; sorting out the good from the bad science takes time. Nevertheless, in this ever-more-speedy environment where information passes as wisdom and the need for public recognition often trumps caution and confirmation, snap scientific observations are increasingly the norm—often with laughable when not outright tragic consequences. What's sorely needed is a long-term method of contemplating the relationship between the brain and the mind that will not be made obsolete by the next study or observation. If 2,500 years haven't given us a coherent view of the mind, it's about time that we consider alternative possibilities.

To put the arc of this book into perspective, imagine that

you are eager to begin a research project, but don't have access to a necessary state-of-the-art microscope. You arrange to borrow one from a friend who owns a pawnshop for high-tech equipment. You have no idea who previously owned the microscope, what condition it is in, or whether or not to trust your friend's word that it "works fine." Like any first-rate scientist beginning an investigation, you would first check out the microscope's lenses to be sure there were no smudges, defects, or peculiarities that might create optical illusions or in any way distort the images. After all, the accuracy of your research will always be limited by the quality of your tools.

Traditionally when we think of scientific tools, we think of machinery and methods of inquiry. It is up to the scientist to ascertain that the equipment is flawless, the methodology is unimpeachable, and the results are transparent and reproducible. But this way of thinking about science overlooks the basic limitation of scientific inquiry: it is our minds that dream up the questions and seek out the answers. If the mind generates our investigations of itself, wouldn't it make sense to understand the potential limits of this tool in the same way we would look for flaws in the optics of a microscope?

My overriding goal in writing this book is to challenge some basic assumptions that permeate the field of brain study—from hard-core neuroscience to experimental cognitive science to the more theoretical arguments posed by philosophy of mind. Though experimental cognitive science and more basic neuroscience are often viewed as separate disciplines operating at different levels of investigation (clinical versus basic science), I will be lumping them together as overlapping methods for thinking about the mind. For simplicity, I will refer to the entire field as neuroscience. The book will draw from a number of scientific studies, but will also rely on personal thought experiments and experiences as well as clinical examples and even literary observations. The book is best read as a late-night meditation. Rather than promising answers to age-old questions about the mind, it is my goal to challenge the underlying assumptions that drive these questions. In the end, this is a book questioning the nature of the questions about the mind that we seem compelled to ask yet are scientifically unable to answer.

I'm not suggesting that these are necessarily the only or the best set of interpretations or approaches to the major problems of the mind. Given my argument that bias and irrationality are often unavoidable aspects of any line of reasoning, it's likely there are flaws in my logic and reading of the works of others. Fortunately, it isn't necessary that every observation be unassailable. If any of what I'm suggesting rings true, that is enough. My goal isn't to present one-stop answers, but rather to highlight potential pitfalls and impasses and to provoke an increased awareness of how inherently difficult the study of the mind will always be. Recognition of where the lenses of our primary tool of investigation are inherently blurred allows us to factor in these distortions. If a tool is incapable of making certain observations, so be it. At the risk of an excess of cynicism, it seems to me that if I have a compelling reason to believe that I can't fly, I would be better off returning my wax wings to the scientists who sold them to me than opening the window and jumping to the wrong conclusion.

Though this book will offer a number of criticisms of modern neuroscience, it is not intended as an indictment of the field or of individual scientists. I have the greatest admiration for the central role neuroscience and neuroscientists play in improving both our daily lives and our self-understanding. There is no alternative to the scientific method for studying the physical world. Many of my arguments will be supported by evidence from basic and cognitive science.

To date, there has been a great deal of public and academic criticism about specific technologies such as fMRI. An fMRI can detect an increase in blood flow to an active brain region. It can give a dynamic picture of changing degrees of brain activation while performing a task—either a thought or an action. But an fMRI can't provide a direct measure of neuronal activity. Still, my primary goal is not to look at methodological difficulties that may be overcome in the future.[7] My interest is in pointing out essential limitations that are unlikely to yield to future technological advances. To this end, I will focus on the limits of what neuroscience can investigate and what it can properly conclude. Also, I am not arguing against speculation; after all, this entire book, though drawing on scientific evidence, is purely speculative. And, in the end, that will be the overarching theme of this book. Scientific method, if properly employed, can produce a wealth of useful information. However, any application of this data to explain the mind will always be a personal vision, not a scientific fact.

1 • The Shape of Your Mind

It's unbelievable how much you don't know about the game you've been playing all your life.

—attributed to Branch Rickey[1]

All complex biological systems—which include you and me—use sensory feedback to monitor their environment. We are made aware of the external world through senses such as sight and sound; we know our interior physical world through internally generated feelings such as hunger and thirst. As the vast majority of thought originates outside of consciousness, it seems reasonable that we would also have evolved a sensory system for informing the conscious mind what cognitive activity is going on subconsciously. Without a method for being aware of this activity, it is hard to imagine what role the conscious mind would have, or even if there would be such a thing as a mind.

If we were cars, our minds would have LED displays telling us what is going on under the hood. But being subtle creatures rather than machines, we have a far more sophisticated system for monitoring subliminal brain activity. Instead of a mental dashboard full of flashing lights, we have evolved an array of cognitive feelings. For simplicity, I've used the phrase "cognitive feelings" to refer to those mental phenomena that aren't normally categorized as emotions or moods, but rather are the type of feeling we associate with thinking. These include such diverse mental states as a sense of knowing, causation, agency, and intention.

To be meaningful, these feelings must bear some relationship to the cognitive activity they are announcing. Just as the feeling of thirst must trigger the desire for drinking fluids, an awareness of a subliminal mental calculation must feel something like a calculation. And here's the rub. Thirst and hunger are readily accepted as arising from our bodies, but feelings about our involuntarily generated subconscious thoughts often feel like deliberate actions of the conscious mind.

Take an example from the world of visual perception. Imagine yourself at a local football game. You are focused on the game and oblivious of the surrounding faces of spectators. Then, while you are shifting your gaze to look at the scoreboard, your visual system subliminally detects a face in the crowd that it recognizes as your old friend Sam. Your visual cortex compares the incoming image of the face with previously stored memories of Sam's face and calculates the probability that the face is Sam's. If the likelihood is high enough, the brain sends the image of the face up into consciousness along with a separate feeling of recognition. You feel as though you consciously assessed the face and determined it was Sam. Depending upon the strength of this feeling of recognition, you will also sense the degree of likelihood that this facial recognition is correct. This might range from a feeling of merely "maybe" or "it could be, but on the other hand . . ." to a sense of utter certainty.

A visual input of a face, though not initially making any conscious impression, has triggered two separate unconscious brain activities. One is exclusively mechanical and without any associated feeling tone—the comparison of Sam's face with all other faces previously stored in memory. The other is purely subjective sensation—the feeling of recognition. The two arrive in consciousness as a unit—the visual perception of Sam's face and the simultaneous feeling that it is indeed Sam. Even though this process occurs outside of awareness, we feel it is the result of the act of conscious recognition. *Such lower-level brain activities are experienced at a higher level as voluntary acts.*

Because we know the brain is superb at subliminal pattern recognition, we find it relatively easy to concede that recognition doesn't take place consciously, despite how it might feel. But there are a host of mental sensations that are so intimately linked with our conception of conscious thought that the idea of them not being under our conscious control seems far-fetched.

In my previous book, *On Being Certain*, I introduced the concept of involuntary mental sensations—spontaneously occurring feelings about our thoughts that are experienced as aspects of conscious thought. Though we feel that they are the result of conscious rumination and represent rational conclusions, they are no more deliberate than feelings of love or anger. My focus was on feelings of knowing, certainty, and conviction—feelings about the quality of our thoughts that range from vague hunches and gut feelings to utter conviction and a profound "aha."

I now realize that feelings of knowing are a small part of a larger mental sensory system that includes the sense of self, the sense of choice, control over one's thoughts and actions, feelings of justice and fairness, and even how we determine causation. Collectively, these involuntary sensations make up much of the experience of having a mind. In addition, they profoundly influence how we conceptualize what a mind "is."

It is vitally important to realize that the cognitive aspect of thought—the calculation—has no feeling tone. Our entire experience of these calculations comes via separate feelings that accompany them into consciousness. For example, though contrary to personal experience, there is no way to objectively determine the origin of a thought. When an idea "occurs to me" or feels as though it "popped into my head," I tend to label it as arising from the unconscious. On the other hand, if I have the feeling of directly thinking a thought, I am likely to conclude that it is the result of conscious deliberation. The distinction between conscious and unconscious thought is nothing more than our experience of involuntary mental sensations.

This separation between thought (the silent mental calculation) and feelings about thoughts is central to any inquiry into what the mind might be. We know the mind only through our experience; it isn't something that can be pinned on a specimen board, weighed or measured, poked or prodded. Seeing how our sense of a mind arises from the messy and often hard-to-describe interaction of disparate involuntary sensations is a necessary first step toward any understanding of what the mind can say about itself.

Our Brains, Ourselves

A car cuts you off on the freeway and you get enraged; you honk, flip the driver the finger, fume, and carry on about this brutish lack of manners being a surefire indication of civilization's impending demise. Your spouse, for the thousandth time, urges you to learn a little self-control. Of course, dear, you halfheartedly agree, your mind oscillating between further thoughts of revenge and the painful recognition that you have just acted like a two-year-old.

You quickly drum up a bevy of seemingly reasonable explanations—a stressful day, a poor night's sleep, the new anti-hypertensive medication you started a couple weeks ago, long-standing control issues and unresolved childhood slights, your growing apprehension over your declining IRA account. On the other hand, your father had a hair trigger and was prone to seemingly unprovoked tirades and furies. Perhaps you inherited some angry strands of tightly wound DNA. If only there was a straightforward method for self-examination. But your mind reels at the seemingly infinite combination of possibilities, as though the very concept of self-awareness is an overrated myth, a low-probability rubber crutch for the emotionally desperate.

Nevertheless, you have to start somewhere. Though changing your genes is presently out of the question, perhaps you can address your financial concerns. Back at home, you review your IRA portfolio. Your best friend, a financial wizard, gives you a thousand reasons why stock prices are at a generational low and insists that you should "buy, buy, buy." His arguments are persuasive. You boot up your online broker and poise your finger over the Buy button; but, as though controlled by invisible forces, you have a complete "change of heart" and sell everything. You are puzzled by your behavior. It is as though you have lost control of yourself.

Later that night, flipping through a popular psychology magazine, you read that fMRI studies have shown that the brain region for controlling hand movements is activated before you are aware of making a decision to move your hand. Brain wave (EEG) studies confirm the finding. This can't be, you think. You try a simple lab experiment. You think about moving your hand but don't make the final decision to move it; your hand rests quietly in your lap, awaiting instructions. You then consciously decide to wiggle your fingers. You exert some effort, and, not surprisingly, your fingers wiggle on command.

But if the fMRI and EEG studies are correct, your experience of wiggling your fingers voluntarily is nothing more than a comforting illusion foisted on you by an unconscious with its own agenda. Only after the fact did your subconscious let "you" know what it had already decided and acted upon. Looking down at your hand as though it has a mind of its own, you wonder who exactly made the decision. "Who am I?" you ask yourself, at the same time wondering who is doing the asking and who is expected to answer.

To come up with anything remotely resembling a tentative answer to what the mind might be and do, we first need some working understanding of the placeholder for the mind—the self. A mind isn't an impersonal organ like a liver or a spleen; it is an integral aspect of a self, a part of what makes us an individual as opposed to an object. It's the center of our being, the main control panel for our thoughts and actions. The self's central function—creating thoughts and actions—is commonly what we mean when we talk of a mind. Evolution isn't a linguist. At a practical level, the mind and the self are inseparable. It is hard to imagine a functional self without a mind, and vice versa. Both are essential constituents of an "I." When a patient with Alzheimer's disease "loses his mind," he is invariably described as having also lost his "selfhood." Though we can readily create thought experiments about a brain in a vat, we cannot even begin to think of a mind in a jar. The mind needs to be physically embodied; there needs to be a something or someone that is having the thoughts and performing the actions. Fortunately, we have a built-in set of mechanisms for creating a visceral sense of a home for the mind—the physical self.

Where Am I?

Perhaps the most universal yet personal of all involuntary mental sensations is the feeling of where "you" exist in your body. Many of us have a very strong sense that the center of our being is a few inches behind our foreheads, just above our eyes. But, if we could take off the top of our skull and closely dissect each of our brains down to the subatomic level, we would find no homunculus, no little "I" running the ship, minding the store, holding the reins of our conscious actions, or even idly ruminating. The center of our being is nowhere to be found.

At the purely intellectual level, even the most science-impaired among us understand that mental states, no matter how seemingly psychological in origin, ultimately arise out of brain states. Everything we experience is generated by mindless brain cells and synapses. Nevertheless, we cannot shake the contrary feeling that there is a personal "I" that is sufficiently separate from these states to have an understanding of this proposition. As I write this statement, I have the undeniable feeling that there is a special "I" that is both writing and reading this sentence, and that this "I" resides within a larger unit over which I claim at least some degree of responsibility—my personal body.

It is impossible to imagine what it would feel like to lack a consistent sense of self. We couldn't take complex actions, consider "what might have happened" in the past, contemplate a future, or make plans.[2] Yes, it would be wonderful if we could occasionally flip a switch that would free us from the downside burden of perpetual self-involvement and internal dialogue, but that's not the way the physical sense of self works. It is as involuntary as hunger and thirst.

For now, put aside considerations of personal aspects of a self—the narrative of your life that you tell yourself and others. I want to focus on the more basic physical sensations that collectively create the scaffolding of a self onto which you hang your personal memories, stories, and experiences, for it is this

physical sense of self that creates the housing for our experience of a mind. We don't experience our mind as being
several blocks away in a saloon, nursing a drink while contemplating the universe. For most of us, most of the time, the
mind resides within our personal sense of the dimensions of
our self.

The spatial qualities of the experience of a mind arise from
subconscious brain mechanisms. Our thoughts about the dimensions of mind aren't similarly constrained by our biology.
Conceptually, the mind can be anything that we imagine.
This distinction between felt experience and theoretical views
of a mind is critical to any larger understanding of what a
mind "is." To lay the groundwork, we first need to know how
our experience of the dimensions of our minds shapes our
investigation of our minds.

A basic tenet of neurological research is to break down a
complex mental state into more manageable subcomponents.
One method—studying patients who have had a discreet brain
insult that clinically affects only one of these subcomponents—
has been instrumental in providing a good practical model of
how sensory systems work. For example, this technique has
helped reveal that the visual system is composed of a number
of specialized neural circuits (modules); each processes an aspect of vision such as lines, edges, color, or motion. Collectively they create a visual image. We will use this same approach
to dissect out the various mental sensations that collectively
create a sense of self.

To begin, listen to these three descriptions of how abnor-
mal electrical brain activity can dramatically alter our sense of
self. As you read these histories, notice how the physical sense
of self can be distinguished from the felt location of your particular first-person vantage point in the world.

CASE HISTORIES

Patient #1

A twenty-one-year-old man with a six-year history of poorly controlled seizures awakened with a peculiar dizzy feeling. He got out of bed but, when turning around, saw himself still lying in bed. He became angry about "this guy who I knew was myself and who would not get up and thus risk being late at work." He tried to wake the sleeping body by shouting at it, then by trying to shake it and then repeatedly jumping on his "alter ego in the bed." The body didn't respond. Puzzled about his apparent double existence, the patient became frightened by his inability to tell which of the two he really was. Several times his bodily awareness switched from the one standing upright to the one lying in the bed. While in the lying-in-bed mode he felt quite awake but completely paralyzed and scared by the figure of himself bending over and beating him. He walked to his bedroom window, looked back to see his body still in bed. "In order to stop the intolerable feeling of being divided in two," he jumped out the third-story window.

Fortunately, he landed in a bush and sustained only cuts and bruises. Neurological evaluation documented recurrent seizure activity triggered by a slow-growing tumor in the man's left temporal lobe. The tumor was successfully removed.[3]

Fiction is filled with fascinating tales of the doppelgänger. In Edgar Allan Poe's "William Wilson," the protagonist, in an attempt to stab his double, kills himself. In Oscar Wilde's *The Picture of Dorian Gray*, the hero commits suicide in order to escape the horror of being haunted by his second self. When reading such stories we readily suspend disbelief; we understand that they are intended as metaphors rather than realistic

Patient #2

A fifty-five-year-old man with a history of seizures from age fourteen developed recurring stereotypical attacks of strange bodily sensations. Without warning he would feel that a stranger had invaded the left side of his body and that the left half of his head, the upper part of his left trunk, and his left arm and leg no longer belonged to him. During an episode, he felt that he existed only in the right side of his body, yet he was able to carry on normal activities, including giving lectures.

accounts. Though the above case history also feels fictitious, it is a very real description of a relatively rare neurological phenomenon—autoscopy—seen with certain types of paroxysmal brain activity (i.e., seizure disorders and migraine). This experience of a transiently divided and unstable sense of self provides an excellent starting point into how the physical self is generated.

Patient #3

A thirty-year-old man had a twenty-year history of seizures characterized by an overwhelming sense of numbness in his legs, chest, and neck. During the attacks he would lose awareness of everything below his chin so that his head felt detached from the rest of his body. He simultaneously experienced himself as an observer of his body and the subject of observation.[4]

In the first example, patient #1 feels and knows that he is standing by the bed, yet continues to have the unshakable feeling that the illusory body in bed belongs to him. This sense of ownership—sometimes referred to as a feeling of "mineness"—is thought by many cognitive scientists to be central to self-awareness. Take a look at your hands. Though you have no doubt that these hands are yours, this determination doesn't require any conscious deliberation. This feeling of ownership is pure sensation, no different from feeling the weight of the book that you're presently holding in your hand. Visual and proprioceptive inputs inform us about the position of our body parts; touch inputs tell us what our body is in contact with. This information is converted into a coordinated body schema—a representational map of the body and its relationship to the outside world that is experienced as "mine."

Imagine our perpetual confusion if we had to first distinguish what is "us" from what constitutes our surroundings, or even worse, who is "me" and who is "another." If you saw a hand rapidly approaching your face, you wouldn't be immediately able to know if you were about to unconsciously scratch an itch or were getting mugged. Mercifully, evolution has given us a built-in method for immediate recognition of our body and its parts. This sense of physical self not only allows us to move through the real world; it also enables us to navigate the world of our imagination—both alternative pasts such as what might have happened "if only . . ." and projected scenarios for the future. Onto this most basic map of the dimensions of "me" we hang our nonphysical self—our collected thoughts and memories.[5]

The feeling of ownership is dependent upon unconscious mechanisms telling us that the current body image closely approximates previously stored body imagery.[6] Patient #2 provides us with a dramatic picture of how this sense of ownership of body parts can be temporarily switched off during a seizure. This loss of sense of ownership of one side of the body is commonly seen in patients with lesions in a region of the right parietal lobe believed to be instrumental in the generation of our body schema. Patients characteristically comment that the affected side of the body no longer feels "theirs"; they will often attribute ownership to an outside force. I still remember

Mrs. A, an elderly deacon in her Pentecostal church, who, after a stroke in the right parietal region, claimed that the left side of her body belonged to the devil. She repeatedly grabbed her paralyzed left arm with her right hand and tried to throw it out of bed. No amount of explanation reassured her that her arm was "hers."

(A word of caution: though denial of body parts is most commonly associated with right parietal lesions, premotor regions of the frontal lobe also have been implicated in the sensation of a sense of self, including recognition of body parts. However, the final word as to the underlying neuroanatomy isn't in.[7] The purpose of this discussion is to point out the involuntary neural basis of such mental states, not to take a stand on the underlying neuroanatomy, as this may well be modified with further studies.)[8]

A second feature common to patients #1 and #3 is an altered sense of where "I" exists. Patient #1 describes a sense of "I" that rapidly shifts back and forth between the two bodies. Patient #3 describes the equivalent of a mental double vision in which he is simultaneously observing and being observed. These descriptions of a short circuit–induced sense of "I" separate from a sense of the physical body underscore its origin in brain wiring.

This dissociation between the felt location of "I" and where your body is physically located is the hallmark description of an out-of-body experience (OBE). Swiss neuroscientist Olaf Blanke and colleagues have shown that an OBE can be directly provoked with seizure activity, brain stimulation, and certain psychoactive chemicals.[9] Patients typically describe floating or hovering above their own bodies, perhaps watching a surgical procedure or even witnessing their own death. Despite the innumerable fanciful explanations, OBEs are nothing more than illusions generated by quirks of physiology.[10]

The investigation of such perceptual illusions has provided some remarkable insights into how easily our sense of self can be physically manipulated. Perhaps the most generally known is a simple parlor trick—the rubber hand illusion. A test subject is seated at a desk, with one hand placed under the desk so that it is hidden from view. On the desk in front of him is an artificial rubber hand. The subject is asked to focus his attention on the rubber hand. When the experimenter simultaneously strokes the hidden and rubber hands with a paintbrush, the visual information—seeing the rubber hand being stroked—overrides the subject's proprioceptive knowledge of the position of his hidden hand. Within a minute or two, the subject perceives the rubber hand as his own.[11]

More recently, the studies have been extended to include the entire body—the body swap illusion. By altering the perspective with which subjects see themselves, the sense of ownership can be transferred to a mannequin or even another person. The trick is to create a visual image of the subject's body that is different from our customary experience of looking down at our bodies from the perspective of our eyes. To do this, subjects are fitted with a head-mounted display—a modified virtual-reality headpiece connected to a video camera placed behind their backs. Seeing themselves from the camera's point of view induces a form of OBE; the subjects describe their center of awareness as being located outside of their physical bodies.

After determining how the sense of location of the self can be readily manipulated by altering sensory inputs, two Swedish neuroscientists have devised several fiendishly clever experiments.[12] In one study, they make subjects feel a sense of ownership of a mannequin standing alongside them. Even though the subjects can clearly see that the mannequin is not them, the subjects cannot shake the feeling that their sense of self is located inside the mannequin. The extent of this altered sense of self can be seen when the mannequin is approached with a knife. Subjects display increased sweating and increased skin conductance response, and may describe feeling anxious. Changing their sensory inputs makes the subjects emotionally identify with an illusion.[13]

The evidence is overwhelming: the most basic aspects of a sense of self—its physical dimensions and where we experience our center of awareness—are constructed from sensory perceptions. It is hard to avoid the comparison with a virtual avatar. Both are pure constructs that nevertheless generate a real sense of personal identity. And, just as the height, weight, or dimensions of an online avatar can be changed, the dimensions of a self are also subject to change. Consider the following monkey-rake experiment.

In the 1990s, using microelectrode recordings, cognitive neurobiologist Atsushi Iriki located neurons in the parietal region of monkeys that respond to both visual and touch input.[14] These cells fire when an object is placed near the monkey's hand, as though announcing the presence of an object within the monkey's reach. The monkeys are then taught to use a rake to extend their reach. Soon the same cells fire whenever the monkey sees anything within reach of the rake. The rake has been incorporated into the neural representation of the monkey's body schema as an extension of its arm and hand.[15]

The speed with which this change in body image mapping occurs is quite impressive. Macaque monkeys rarely use tools in the wild but can become proficient in tool usage with a few weeks of training. Within a week of beginning training, fMRI scans will show increased volume of gray matter in the same regions that exhibit increased neuronal firing rates.[16] Though the exact structural explanation for this increase in brain volume remains unclear, it is thought that this may represent either new blood vessel formation in the region or even possibly the generation of new brain cells (neurogenesis).[17]

Whatever the underlying anatomic explanation, it is clear that the brains of these monkeys can easily be rewired by the use of tools. The same seems likely to apply to us. In 2009, fMRI studies showed similar areas of brain activation when volunteers were exposed to new tools.[18] Though I am generally suspicious of pat evolutionary explanations for all physical adaptations, it is hard to avoid the conclusion that evolution has provided us with a certain innate flexibility of body image that has allowed us to acquire the skillful use of tools. But this mutability of our representational maps of our body extends beyond tools.

Seventy-three-year-old Mrs. B sustained a large right hemisphere stroke resulting in left arm paralysis. She remained cognitively intact with no signs of confusion, yet demonstrated a total lack of awareness and sense of ownership of her paralyzed left arm, repeatedly claiming that the arm belonged to someone else. Surprisingly, this lack of any feeling of ownership included her wedding rings, which she wore on her left hand. Though she was able to clearly see and describe the rings, she denied that they belonged to her. When the rings were shifted to her right hand and then shown to her, she immediately felt they were hers. To see if this loss of ownership extended to other objects not normally associated with her left hand, a comb and a set of keys were placed in her paralyzed hand. Both were recognized immediately as "my comb" and "my key-holder." The lack of ownership was sharply circumscribed, limited to that set of objects—her wedding rings—that were historically related to her sense of her left hand. It was as if, before the stroke, the rings had been included in an extended, primarily visual body schema.[19]

Pundits are quick to point out that web surfing, video

games, online virtual environments, Twitter, Facebook, and innumerable other technological innovations are changing our brain circuitry. The notion of an extended mind has become commonplace, with the monkey-rake experiment frequently being offered as hard evidence for how these changes take place. Yet even as we acknowledge the profound effects of our environment on our brain, most of us continue to feel that we possess our own individual minds that are at least partially immune from insidious outside influences. It is hard to overcome the basic sense that a mind is embodied within the confines of a self—the essential distinction between the experience of a mind and the concept of a mind.

I recently heard an eminent philosopher and logician, the University of California, Berkeley, professor John Searle, say, "The extended mind ought to seem mistaken on the face of it. . . . In general, in philosophy, if you get a crazy result, it's false. . . . If someone gives you an account of the mind that runs counter to your own experience, well then, you know they made a mistake."[20] Searle is well aware that all experience is subjective perception, and all perception is filtered through undetectable biases and predispositions. Personal feelings shouldn't be the sole measure of an idea. Nevertheless, Searle cannot step back from his own feeling that he possesses a unique mind (the sense of ownership) residing within his personal body (the sense of physical self) that has willfully drawn this reasonable conclusion (sense of agency). This combination of involuntary mental sensations restricts his ability to entertain alternative possibilities for what a mind might be—a poignant reminder of how philosophical conclusions can be guided by biology rather than by reason.

The description of Mrs. B provides a Rube Goldberg–like insight into how this conflation between the sense of a self and a mind might affect the very experience of our thoughts.

(For the purpose of this discussion, I am separating the mind from a sense of self, though obviously there is a significant degree of overlap and neither seems capable of an independent existence.) Over time Mrs. B's sense of ownership of a body part (her hand) came to include the body part's contents (her rings). As the mind is a lifelong key component of the self, a sense of ownership of the self similarly is likely to extend to its contents—the mind. In turn, as our thoughts are a central component of our mind, we might expect some overflow of this sense of ownership from the mind to its contents—our thoughts. Each of us possesses "my mind" and "my thoughts."

As we explore the role of mental sensations in creating the structure of the self, try to get a sense of how you know when a thought is "yours." Whether this feeling of having your own thoughts is a willful determination or an involuntary mental sensation is critical to how we determine what a mind "does."

Moat Around the Castle

Before leaving the subject of the physical dimensions of a self, we should take a quick look at a separate but closely allied phenomenon—the sense of a personal space. Each of us has his own preferred distance between himself and others. Move closer and you're "invading my space." If the physical sense of self is the brain's version of a car's Global Positioning System (GPS), which tells you where the car is in relationship with the environment, the sense of a personal space would be equivalent to a car's sensors, which tell you when you are too close to another car or about to back into a telephone pole.

This sense of personal space can be dramatically altered with brain disorders.[21] The most extensively studied example is S.M., a woman in her midforties with a rare genetic condition that has resulted in extensive bilateral damage to her amygdala.[22] S.M. has a reputation for being fearless and superfriendly, with a tendency to "violate" what others perceive as their own personal space.[23] During a series of trials exploring her sense of personal space, she admitted that she didn't feel discomfort at any distance from the experimenter. On one occasion, she walked up to the experimenter until their noses were nearly touching. The experimenter was the only one who felt uncomfortable.[24]

A similar lack of need for a personal space is seen in another genetic condition—Williams syndrome. Though saddled with a number of developmental problems and learning disabilities, patients with Williams syndrome are highly social and friendly, and are typically unafraid of strangers.[25] Like S.M., they apparently don't feel the need for any significant degree of personal space; as caretakers will attest, they are commonly "in your face."

Both S.M. and children with Williams syndrome exemplify the role of basic brain mechanisms in determining the size of the buffer zone with which we unconsciously surround our physical sense of self. I'm not suggesting that this feeling is exclusively shaped by our biology. For example, ethnic and cultural influences clearly make a difference. Those from the Middle East and southern Europe seem comfortable with less interpersonal space than northern Europeans.[26] Whether a feeling arises exclusively out of innate biological factors, sociocultural influences, or any combination of nature and nurture, the final result will be expressed as a mental sensation.

That a sense of personal space is likely to be the product of separate brain mechanisms than the sense of a physical self should serve as a huge caveat. Though this and subsequent chapters are designed to show how a variety of cognitive feelings shape our sense of self, the list is by no means complete. I suspect that cognitive scientists and/or you readers will come up with suggestions for other potential candidates. Also, it is likely that these various sensations will be experienced and described differently. For example, I recently asked a group of psychoanalysts where they felt the central core of the "self" was located. Most opted for the head, while one senior clinician rubbed her upper abdomen. When I asked her why she picked her abdomen, she said with a wry smile, "Because that's how I feel." This discussion isn't meant as a written-in-stone compendium of specific sensations so much as a way of understanding how the physical self is a pure projection of involuntary sensations.

A second problem to briefly consider: if the physical self is pure sensory projection, is it correct to call it an illusion? On the one hand, the physical sense of self is clearly an illusion in that there is no underlying entity to which the feeling refers. On the other hand, it is important to avoid thinking of illusion pejoratively—that somehow an illusion isn't "real." The sense of the self, even if we call it an illusion, is as real as pain, anguish, or love. (We will later address this problem of "real" versus imagined or psychological—a false dichotomy commonly seen in medical reporting.) Despite my ambivalence over the term, I have decided to go with "illusion" for quite practical reasons. First, understanding the nature of an illusion is critical to dealing with it. If I see a straight glass rod bending as it is immersed in a beaker of water, I can make appropriate adjustments in how I think about what I'm seeing. Second, it is possible that contemplating the illusory nature of the physical self can lead to constructive therapeutic options—ways of actually altering how we experience our physical self.

For instance, seeing the physical sense of self as an illusion provides a potentially enlightening perspective on some otherwise puzzling mental disorders. Several years ago, I watched with utter horror and disbelief a Discovery Channel documentary on patients who wished to have a perfectly functioning arm or leg amputated. First described in 1977, this condition, variously referred to as apotemnophilia or body integration identity disorder, has generally been thought to be due to deep-seated psychological forces. After all, anyone who wants a good arm or leg cut off must be crazy. But the case histories of many of these patients fail to show any hint of what the underlying reasons might be. Often there are no associated psychological quirks or symptoms, nor is there any prior psychiatric history. One sufferer describes being only four or five when he first became fascinated at the sight of an amputee. He recalls that by age seven he was standing by a bus and said to himself, "If I just stick my leg under the rear wheel of the bus, it will run over it and my leg will have to get cut off."[27]

In 2009, behavioral neurologist V. S. Ramachandran and colleagues studied four men with this disorder. Using a form of functional imaging, magnetoencephalography, they were able to detect a striking reduction (compared to normal controls) of electrical activity in the right superior parietal lobule when the affected limb was touched.[28] The area of decreased activity corresponded to the area that was affected in Mrs. A, who thought her arm was the devil and was preoccupied with throwing it out of her bed.

Ramachandran has postulated that there is a primary neurological malfunction in this region of the right parietal lobe that coordinates incoming visual, sensory, and motor inputs to produce a dynamic body image. Knowing that this circuitry is defective doesn't tell us the cause of this defect, but it does offer yet another view on how some mental conditions might arise from a disturbed physical sense of self. It also raises the possibility of new ways of contemplating therapeutic interventions.

Using the same principles underlying the rubber hand illusion, Ramachandran has developed the mirror box illusion to effectively reduce phantom limb discomfort in amputees.[29] The principal is straightforward: you try to trick your brain into replacing the image of the missing part with an image of a perfectly functional normal extremity. If it's an arm that's amputated, the patient positions a mirror in his line of sight in such a way that when he looks at the missing arm, he sees only his good functioning arm. Various techniques such as gentle exercise or lightly stroking the hand augment this visual input. Functional imaging studies have shown that the brain undergoes reorganization of the image of the missing arm. The greater the degree of reorganization that occurs, the greater the degree of reduction in phantom limb pain.[30]

The same effect can be produced with a virtual environment in which an avatar's hands are positioned so that the subject feels that the avatar's hands are his own. By transferring the feeling of ownership to a virtual hand, recent amputees find it easier to adapt to a prosthetic limb. In effect, the patient's brain is reprogrammed to accept the prosthetic limb as his own.[31] Another example of the benefit of altering the body image is the use of lenses (inverted binoculars) to make a hand look smaller than normal. Patients with chronic hand pain who observed their affected hand as smaller had a significant reduction in their pain. Even more fascinating, these patients were noted to have measurably less movement-induced swelling. Conversely, when using binoculars to magnify the size of affected hand, the degree of reported pain and measurable swelling was increased.[32]

Intentionally altering a body's representational maps could have far-reaching implications in a variety of disorders characterized by faulty body imagery—from anorexia nervosa to the body dysmorphic disorders that result in excessive plastic surgery (think of Michael Jackson or Joan Rivers).[33] At the very least, such demonstrations highlight the intimate relationship between altered body imagery and how we see ourselves in the world—a perspective integral to scientific inquiry.

In summary, a sense of a coherent body image, where our center of awareness is located, and our first-person perspective on the world collectively establish the general frame of a self. We can even toss in the metaphor of a protective moat (sense of personal space) that surrounds this castle that is home to our personal kingdom. However you describe these feelings, what seems inescapable is that pure sensation is the sole necessary and sufficient prerequisite for the physical sense of a self onto which all of us, neuroscientists and philosophers included, hang our perceptions and generate our ideas of a mind.

2 • Agency, Will, and Intention

The play dictated itself, but I confess that I wrote it—with intent, maliciously, purposefully, in command of its growth.

—Harold Pinter, Various Voices: Prose, Poetry, Politics, 1948–1998

It's a stormy night. You are driving on a narrow two-lane road. Just as you round a bend, a tree suddenly falls directly across your lane. Without thinking, you swerve across the double line, avoiding the tree but smashing into an oncoming brand-new BMW. Fortunately, no one is hurt. You apologize to the other driver, then blurt out that you're not a bad driver and that the accident was unavoidable. "It was pure reflex; I didn't have time to think." Totally preoccupied with how this collision will jack up your insurance premiums, you absentmindedly reach into your glove compartment, pull out your registration and insurance documents, and hand them to the other driver.

In this scenario, you have assigned two actions quite different degrees of intention and willfulness based upon how the actions felt. Despite performing both actions without any conscious awareness, you attributed your loss of control of your car to mere reflexes, while taking full responsibility for handing the other driver your papers. How we think about ourselves, even how we determine what represents "me" as opposed to mere biology, is intimately related to how we understand mental states such as feelings of effort, will, intention, even "what I'm doing now."

Putting the Self into Action

Watch your dog or a friend run after a Frisbee that you've thrown. Neither runs to where the Frisbee is presently located; both subliminally calculate what would be the optimal intersection between his present running direction and the Frisbee's flight path. The same is true when an eagle catches a sparrow in midflight, or a lion scurries for his gazelle snack. Successful motor behavior is dependent upon continuously updated subconscious calculations that take into account everything from wind velocity to the heaviness of the air to the firmness of the terrain underfoot. To do this, the brain has to project a "self" situated in and capable of making its way in the external world. Maps of the "self" and of the external world, and templates for motor actions, must be in place in order for your brain to imagine "you" running in various directions at a variety of speeds and then to calculate the optimal path for you to take to catch the Frisbee.

New York University neuroscientist Rodolfo Llinás has gone so far as to suggest that planning and predicting motor movements is the primary reason for having any mental life.[1] Though this might sound too simplistic, his point bears consideration. All behavior is the execution of some motor plan. The most abstract of thoughts is the contemplation of some past or future action. It is impossible to think about any issue, from voting to meditating, without considering what you will be "doing." Even doing nothing is an action. Thinking is the motor activity of a mind.

Llinás points out that brains are limited to organisms that move. A tree has no need of a central nervous system because it's not going anywhere, but an animal on the prowl needs to see where it's headed and needs to predict—perhaps even envision—its future place in the world. As poster child for his theory, Llinás offers the sea squirt. This marine creature starts life as a motile larva, equipped with a rudimentary brainlike collection of about three hundred neurons. But when it finds a hospitable site on the ocean floor and puts down roots, it stays put. Without the need to wander, it apparently has no further use for its brain, and so it eats it.

Fortunately for those of us without a strong stomach, we have evolved beyond the sea squirt; our minds are always on the move even when we are standing still. Once this "self" is up and running, it needs to be propelled by purpose and the feeling of being in control of mental as well as motor actions. Without a palpable sense of controlled and purposeful behavior, all would be reflex; there would be no need for a mind to consciously make any decisions.[2]

Welcome to the sense of agency—the feeling that you are the one who is causing or generating an action. To make a few necessary distinctions, let's drop in on your Friday-night poker game. The rube to your right has bet and it is up to you to raise, call, or fold. You look down at your hand, deliberate on whether to toss in some chips or throw in the towel. Seeing your fingers poised over your chips, you are fully aware that they are your fingers (sense of ownership). You are actively and deliberately reaching for your chips (sense of agency). You also are aware of the possibilities of using the chips to raise or call. When you pick up additional chips to raise the idiot who has bet with absolutely nothing, you have the added sense of having consciously decided between the available options (sense of choice).

From the point of view of brain-generated perceptions, a virtual "self" has felt itself making a clever assessment of another's behavior, followed by a conscious decision and a fully deliberate motor action. Nevertheless, these experiences—the senses of self, ownership, choice, and agency—are all components of an involuntary mental sensory system. Collectively they create the experience of a "self in action."

Lift up your arm. If you pay close attention, you will feel that it is your arm that is rising. You will "feel" that you fully intended to lift the arm and are deliberately lifting it. At the same time, the likelihood is high that you will have noticed none of the particulars of this movement. But if your arm is lifted by someone else—for example, your doctor while doing a routine exam—you will feel his grip on your arm, the passive elevation overhead, even a sense of awkwardness of the position of the joints in relation to each other. You still feel a sense of ownership—the arm being lifted is yours—but you experience no sense of control over the movement. You have no sense of agency. Similarly, when struck by your friendly neurologist's reflex hammer, you note the specifics of the knee jerk, but the movement feels strange, otherworldly, not of your doing.

Predicting Agency

As far back as the 1860s, physician and physicist Hermann von Helmholtz was aware that some actions, such as moving your eyes back and forth while staring at an image, cause perceptual problems. There is nothing in the retinal signal that can distinguish between the perception of the image moving as a result of eye movement or actual movement of the object in the external world. What your eyes see is insufficient to determine whether a Frisbee is zipping by at high speed, or is actually hovering motionless in front of you, with the illusion of the Frisbee's movement being created by your rapidly scanning eyes. It's only by having past knowledge of what a Frisbee flight path looks like and the knowledge that you are out playing Frisbee with a friend that allows you to unequivocally "see" the Frisbee in flight. This visual perception is made by the brain, not the eye, and is the result of the brain using both visual input and prior knowledge to calculate the likelihood that the Frisbee is actually in motion. (A similar visual ambiguity occurs when you try to decide whether it's your train or the train on an adjacent track that is pulling out of the station.)

Foreshadowing modern neurophysiology and the understanding of neural networks and feedback loops, Helmholtz suggested that proper perception is based upon prediction. In recent years, neurophysiologists have confirmed the role of prediction, not only in guiding perception, but also in coordinating motor acts. Perhaps the most convincing evidence for a metaphoric "central predictor" is the temporal relationship between activation of the primary motor cortex, which generates a movement, and the "upstream" areas (higher-level cortical regions) instrumental in predicting this action.[3] It has been consistently shown that the activation of upstream prediction occurs long before the motor impulses leave the brain to activate our muscles.[4] The brain calculates where we are to run in order to intercept the Frisbee in midflight; then it sends appropriate messages to our muscles. This is an ongoing process, with the "central predictor" continuously monitoring our direction and speed in order to fine-tune our path. The best physiological explanation is that the intention to activate muscle fibers also fires up a separate feedback loop that informs this "central predictor" of what muscle action is about to occur. In short, we perceive our motor actions both on the basis of sensory inputs from the muscles, tendons, and joints during the action, and via a separate central brain mechanism (a representational map) that has advance knowledge of what we are about to do.[5]

These two independent ways of knowing about a motor

action can be readily demonstrated in patients with disorders
of the peripheral nervous system that block sensory inputs
from their arms and/or legs. Even though such patients have
great difficulty knowing how much force they've exerted with
an affected limb, they retain an accurate sense of how much
effort is necessary for a movement.[6] For instance, if they reach
for a glass of water, they have a sense of the necessary effort
required despite having no sensory feedback from the arm to
tell them how much effort is being exerted. Over time, some
patients are able to use this central sense of effort to gain some
partial control over the extremity stripped of peripheral sensation. Though this sounds suspiciously like flying blindfolded,
even the most clumsy performance of an act without sensory
feedback is conceivable only because our brain is a superb predictor of what we are about to do, how long a particular action
will take, where our body will end up, what the movement
should feel like, and how much force is necessary.

That we have brain maps for future actions shouldn't be
surprising; complex movements would be impossible if there
were not some preceding neural activity to guide the action.
Think of playing the piano. The intention to play a particular
sequence of notes with a certain rhythm and fingering must
be in place before starting to play. We practice endlessly in
order to convert this set of intentions into neural circuitry that
can act independently of moment-by-moment awareness.
While playing the piece, as with most motor actions, we are
mainly aware of what we intend to do. When an action is going well, we are satisfied with knowing that we intended the
action and are in control of it. This sense of agency is enough.
No details of bodily movement are necessary.

If you wanted to construct a system that would have excellent
advance knowledge of what was expected, and yet keep this information out of sight, you might devise a method for suppress-

ing incoming information *when this information corresponds to what is already expected*. You would only want to be notified of unanticipated movements which indicate that something is going awry. For example, sensing whether or not you have intentionally positioned your arm behind your back is the only way that you would know whether you are in the process of scratching your back or being mugged. If you unconsciously anticipated reaching behind your back, there's no reason to be aware of the individual components of the movement. It is enough to know that the action corresponds to what was already expected.

Rewarding Prediction

An update of the century-old Pavlovian drooling-dog experiments has shown how "central prediction" is created. In Pavlov's original experiments dogs were trained to recognize that a bell would be followed by food. Once conditioned, they would salivate on hearing the bell but before the food was delivered. This response was generated by dopamine-secreting cells in midbrain reward centers. Until recently it was believed that these cells function exclusively to provide a sense of reward. Newer studies of monkeys suggest an additional possibility—that the system also signals that there is an error in our prediction about a reward.

Electrodes were implanted into dopamine-secreting neurons in a monkey's reward center. The monkey was presented with a flash of light followed one second later by a squirt of fruit juice into his mouth. Initially the monkey's dopamine-secreting neurons behaved like reward cells; they responded to the juice with increased activity. However, after a period of training, these cells stopped responding to the juice, and instead responded immediately after the monkey saw the light flash but before the juice arrived. They switched roles from delivering a reward to making a prediction—that the juice should be arriving in a second. If the juice arrived as expected, there was no need for any further notification. Things were going as expected.

But if you flashed the light and then withheld the juice, these cells fired less than before the light was flashed—*a diminished response rate*—precisely at the time that the juice should have arrived. In effect, the decreased firing rate was a notification that the expectation wasn't met—the goods weren't delivered. Over time, the relative firing rates with the flash of light and the delivery of the juice are determined by the ongoing accuracy of the prediction. Keep delivering the juice and the cells continue firing with the flash of light. Withhold the juice and eventually the cells no longer respond to the flash of light.

Because the accuracy of prediction can determine the rate of neuronal firing, we have a straightforward way of understanding why incoming sensory information is suppressed when things are proceeding as expected. An optimally efficient brain should send into consciousness only information that needs to be acted on. If nothing is amiss, there's no need to be aware of the details of an action. We need only know that things are going smoothly. Feeling "in control" is our brain's way of telling us that an action is going according to expectation. This sense of agency is analogous to other feelings we have discussed: the feeling of recognition is a notification of subliminal pattern recognition; the feeling of certainty is a reflection of an unconscious calculation of being correct.

University College London neuropsychologists Susan Blakemore, Daniel Wolpert, and Chris Frith have taken an ingenious approach to exploring the roots of the sense of agency. They began by dissecting out a seemingly trivial observation: we cannot tickle ourselves.[7] Their initial theory: our brain knows in advance what we are going to feel; it already has this information available when it sends the motor commands to the fingers attempting to cause the tickling sensation. In essence, by having a preliminary knowledge of what we will shortly be experiencing, we can't be surprised into being tickled.[8]

To test their hypothesis, Frith and colleagues looked at a group that does report the ability to tickle themselves—certain schizophrenic patients: not all schizophrenic patients, just those with the cardinal symptom of believing that they are not in control of their own actions. These patients commonly claim that they are being manipulated by outside forces. I still recall one patient's harrowing description of being a puppet manipulated by cosmic strings. From a patient's perspective, the logic is impeccable. If you believe in cause and effect, but don't feel that you are generating an action, then someone or something other than yourself must be doing it.

Using fMRI scans, Frith and others have shown that schizophrenic patients who lack a sense of agency over their actions have significantly reduced levels of activity in brain regions known to suppress incoming sensory information from bodily movements. (Sorry for the double negative, but reduced levels of suppression is the equivalent of increased sensory input.) Frith has written that the increased awareness of the sensations of their motor movements is associated with a decreased ability to predict the consequences of their movements. In turn, this leads to a feeling of not being fully in control of their actions."[9] Though it is impossible to imagine sneaking up on one's self, such patients are fully capable of just such a surprise self-attack, being able to tickle themselves even when they know what is coming and that it is at their own hand.

It is highly unlikely that neuroscientists will be able to pin down the sense of agency to any specific brain region or set of neural connections. The feeling of agency arises from a variety of sensory inputs; is modulated by yet other areas that control and/or suppress this information; and may be affected by other mental sensations such as the sense of effort and ownership. As a consequence, widely separated brain regions have been implicated, from the frontal lobe to the occipital lobe and the cerebellum.[10] Rather than being a discrete circuitry, the sense of agency is best seen as a widely distributed complex cognitive system that we experience as the feeling of being in control of an intention to perform an action. By operating in tandem with the physical sense of self, the sense of agency creates the feeling that each of us is a willful agent.

Implicit in Frith's studies on suppression of incoming sensory information is the primary role of intention. As Frith has said, "Most of the time you are not aware of what you are doing. What you are aware of is what you *intend* to do. As long as your intentions are fulfilled, you are not aware what movement you are actually making."[11] Consider a jazz improvisation. What you are aware of is your intention to play softly or loudly or with a particular feeling. You are not aware of the position of each finger at every instant. Skillful execution of one's intentions without any bodily awareness is the pinnacle of being "in the zone."

Hand-to-Hand Combat—a Battle of Intentions

Remember the movie *Dr. Strangelove*? Peter Sellers played the mad scientist (Edward Teller parody) with the ominous-looking leather-gloved hand with a seeming mind of its own, alternately initiating a Nazi salute and trying to choke the doctor.[12] Central to one of the great satirical scenes in modern cinema, this loss of control over the actions of a hand is, in neurological jargon, referred to as the anarchic hand syndrome.[13] It can be seen when the two cerebral hemispheres are surgically separated in order to treat uncontrolled seizures, and with injuries to a variety of brain regions, most commonly the frontal lobe on the side opposite to the affected hand.[14] Though the exact location of the injury varies in different case reports, the common thread is damage to the supplementary motor area thought to be responsible for converting intention into self-initiated actions, and that is involved in the selection of what movement to make.[15]

Descriptions of anarchic hand give us a sense of how brain insults can dramatically alter a person's experience of willful control over her actions. One patient's left hand "would tenaciously grope for and grasp any nearby object, pick and pull at her clothes, and even grasp her throat during sleep." To prevent this, the patient slept with her arm tied to the bed; she acknowledged that the hand was hers, but said it felt like an "autonomous entity."[16] Another patient said that "he turned the pages of the book with one hand while the other tried to close it; with his right hand, he tried to soap a washcloth while the left hand kept putting the soap back in the dish; he tried to open a closet with the right hand while the left one closed it." Though these patients recognized the hand as being theirs, they did not feel as if they were controlling it or causing it to move. One patient felt that someone from the moon was controlling her hand. This commonly described assignment of agency to outside forces has led to the syndrome sometimes being referred to as the alien hand syndrome.[17]

Though contrary to the conscious wishes of the patient, anarchic movements are generally complex and purposeful as opposed to nonpurposeful hand movements (motion evoked with direct brain stimulation or the purely reflexive knee jerk response on a physical exam). One particularly revealing case report underscores the problem of assigning intention based upon what we consciously experience as an intention: "While playing checkers, the left hand made a move he did not wish to make, and he corrected the move with the right hand; however, the left hand, to the patient's frustration, repeated the false move." It is impossible to know why the left hand was performing these actions, and the patient, devoid of any sense of having consciously willed the action, could not tell us. But it doesn't make sense to consider these moves as entirely random and unintentional. The anarchic hand was operating according to some specific motor intentions and instructions even if the intentions and instructions weren't consciously known to the patient. The checker example raises the intriguing possibility that the conflicting moves of the two hands represent two choices that the patient was unconsciously considering, but with only one of these actions being associated with a sense of agency. Because of a neurological glitch, we are given a peek at the path not taken.

This real-life demonstration of "he was of two minds" provides a dramatic insight into the separate nature and experience of intention versus willful action. Intention to perform an act is independent of the feeling of having consciously willed the action. Intention can exist subconsciously without having any conscious correlate (as can be seen in the checker player's subconscious intention to make a countermove that was contrary to his conscious intention).

It is only a short step from agency to choice. In the checker player example, they are two aspects of a single act. The patient's sense of control over his right hand's movement was joined by a separate but simultaneous sense that he chose this particular checker move out of all the ones available to him. By contrast, neither feeling was present when he made a countermove with his left hand. The same is true in the earlier poker example. Your mind makes a choice and your hand picks up the chips.

Agency and choice are two sides of the same hand. Both can be seen as feelings of control over a component of oneself—

body (agency) and the mind (choice). So, for purely mental acts, such as choosing a hamburger over a tofu salad, we use the term "choice." If you actually pick up the hamburger at the counter, there would be a superimposed sense of agency.

Fortunately, few if any of you readers will have the direct experience of the anarchic hand syndrome. However, some of you have either witnessed or experienced being hypnotized. Though the underlying physiology of hypnosis remains controversial, it does provide another perspective on the disconnect between intention and sense of agency. A standard demonstration of hypnosis is to ask a subject to raise his arm overhead and also to instruct him not to remember having made the movement. Upon awakening and seeing his arm dangling overhead, the subject might laugh, seem genuinely puzzled, but, most important, claim no responsibility for this action. Agency has flown out the hypnotic window. Or, by hypnotizing the subject into believing that an object is too heavy to be lifted, you can watch him struggle unsuccessfully to exert sufficient strength to lift a feather or a marble.

Given the considerable evidence that hypnosis is a real phenomenon as opposed to "fakery," these parlor tricks provide an insight into the power of suggestion to affect/alter the most fundamental aspects of a sense of self. Though much of the material I've presented so far has focused on the biological substrate of the experience of agency and choice, hypnosis stands as a reminder that the most basic mechanisms for generating a sense of self can be readily manipulated by outside influences.

The physical sense of self provides the mental constructs necessary to experience the dimensions of the self. Add in the senses of intention, agency, effort, and choice, and you have a rudimentary "self in action." They are the bedrock biologic mechanisms necessary for each of us to experience him- or herself as a physically bounded willful agent.[18]

3 • Causation

I am behind the scenes at a carnival, watching jugglers and sword swallowers practicing. Hearing a loud commotion, I turn to see two clowns violently arguing. Clown A is clearly furious, his voice tremulous with rage; clown B has a wide-eyed look of utter incomprehension. A suddenly lunges at B and punches him in the face. B falls down, his feet in the air, his red nose askew. One of the jugglers comes up and asks me what happened. I explain that clown A knocked down clown B. The juggler laughs, turns to the clowns, and says, "Nice work, guys. You fooled the neurologist." Clown B springs to his feet, and the two men start their routine again.

They had been rehearsing. The punch had been a well-planned near miss that looked like the real thing. Clown B didn't fall down because clown A hit him; they had an agreed-upon script that told clown B to fall down. But the script didn't cause B to fall either. Clown B caused clown B to fall. Or so clown B believes. But then, depending upon your belief in free will, maybe clown B didn't cause his own downfall. Maybe his perfect theatrical dive was predetermined at a subatomic level.

Causes' Effects

Perhaps the one mental sensation most responsible for commonplace misunderstanding about the mind is the complex and philosophically vexing sense of causation—how we can know the bedrock reason(s) behind a sequence of events. From a scientific perspective, we have developed a variety of methods for assigning causation. However, at a personal level, we experience causation as a cognitive feeling. Nearly three hundred years ago philosopher David Hume argued that causation is a sensation arising out of our prior experience and inbuilt mechanisms for connecting separate events into a cause-and-effect narrative. It is a shame that we cannot reach back through history and send a note of thanks to those who've provided true insights that have stood the test of time. For what it's worth, thank you, David. In this chapter I will try to see Hume's observations through the lens of modern neurophysiology.

With the clown example, I have no independent way of determining the actual cause of clown B's fall. Before getting the juggler's input, I can only rely on what I have previously learned about physical forces and the mechanics of a punch. Without this knowledge of the nitty-gritty details of how a knockdown can occur, I would be stumped, unable to draw any conclusion as to what caused what. All I would see would be the apparently random unfolding of a series of unconnected events. Prior knowledge is critical to any determination of causation.

Of course, knowing that a punch can knock someone down doesn't guarantee that it actually caused clown B to fall. I'm still dependent upon what my eyes tell me, and as we must grudgingly acknowledge, perception is not the real thing. What I see isn't necessarily what happened. If I acknowledge that my perception might be unreliable, I might mitigate faulty conclusions by first conjuring up all known ways in which perception can go awry. But a complete understanding of potential perceptual errors is dependent on freedom from these same misperceptions. This circularity of reasoning is perceptual bias's omnipresent handmaiden.

To see how we determine causation in the most elementary of situations, consider the following universal experience: you trip over a loose floorboard on your back stairs and stub your toe. Almost immediately you experience pain in the toe. There is no doubt in your mind that your stubbing your toe caused the pain. But this seemingly obvious cause-and-effect relationship between injury and pain wasn't always obvious. We've all seen a young child reach out to touch a hot stove, blithely unaware of potential consequences. "Hot stove causes pain" is a lesson that must be learned from experience. Once burned, twice shy.

Fortunately, we quickly establish a mental representation of the relationship between injury, tissue damage, and pain. Over time, this set of relationships constituting the insight, "Hot stove causes pain," bonds with other bodily-injury experiences to become the embedded generalization "Injury causes pain." Hume's famous problem with induction—that we can never know whether the future will correspond with the past—isn't an issue at the brain level. The brain, having never taken a course in philosophy, is the ultimate pragmatist; what is true is what works. Like any successful oddsmaker, the brain is a predictor of probabilities, not a stickler for the perfect answer. It is sufficient to know the general rule of thumb that the more frequently B follows A, the more likely it is that A causes B. Just as the feeling of recognition is the involuntary sensation announcing a good match between a perceived image and a stored image, causation is the involuntary sensation that arises out of the subliminal prediction that B is likely to follow from A. The feeling of causation begins with a good fit between present events and prior representational maps.

What happens with a less-than-perfect fit? After all, no two experiences are exactly the same. For starters, change the temporal factors in the above example. What if the toe pain doesn't occur immediately but starts two days later? A week later? A month later? As time passes, the sense of causation diminishes. On the other hand, what if the toe pain begins one month after you've tripped over your nasty neighbor's rotten no-good son's skateboard inconsiderately if not maliciously abandoned in your driveway? How you subliminally feel about the neighbor and his son are critical in deciding whether or not the toe pain is due to your modestly elevated level of uric acid, which is causing an attack of gout, or is reason to hire a personal injury attorney who specializes in revenge. Any calculation of causation is rife with subliminal biases.[1]

Agency and Causation

A second factor key to the feeling of causation is the sense of agency. In order to assign causation to any event, we have to believe that A is capable of affecting B. Consider the collision of two billiard balls. To believe that ball A can cause ball B to move, you need to feel that ball A possesses some ability to affect the position of the other ball. This latter requirement has played havoc with philosophers who wonder how a nonmaterial entity—the mind—can have causal powers in a physical world. Nevertheless, at the experiential level, the feeling of agency is sufficient to generate a sense of the mind's causal powers. Combine seeing thought B consistently follow from thought A (pattern recognition) and the visceral feeling that our thoughts have real effects (agency) and we are likely to conclude that thought A caused thought B.

The assignment of agency isn't restricted to the animate and willful. Who hasn't at least once assigned some devious sense of agency to a refrigerator that "refuses to make ice," or gotten mad at a cold as though the cold virus singled him out? Perhaps it is only in jest, but the very question "Why me?" presupposes that there is intention and agency in whatever causes an illness or bit of bad fortune. Ask yourself if you've ever cursed Mother Nature when it rained on your picnic. Intelligent-design proponents assign intention to evolution— evolution caused XYZ because of a grand design by an intelligent superagent. (Note how sensing agency quickly morphs into causation.) Even those most opposed to such faith-based attributions of agency commonly assign agency to abstract ideas. In his recent book *The Grand Design,* Stephen Hawking has said, "Because there is a law such as gravity, the universe can and will create itself from nothing."[2] If we are likely to attribute agency to a refrigerator, a cold virus, and the universe, assigning agency to the mind is a no-brainer.

To flesh out the complex and overlapping relationships between agency, intention, and causation, consider how, as little kids, we learned to establish causation by witnessing the consequences of our own actions. Suppose that I am tempted to flip the shiny On-Off switch on Dad's stereo. I want to know what will happen. When I do, the music comes on far louder than I expected. Mom sticks her fingers in her ears and screams, "No dessert for you." In my room I review the unfortunate series of events. I flipped the switch (agency) with my hand (ownership of body part) *because* I was curious and wanted to see what would happen (intention). My intention caused Mom to cancel dessert. The general principle: The closer the fit between intention and outcome, the more likely we are to conclude that the outcome was caused by our intention.

Personality traits are clearly a combination of nature and nurture, with genes playing a significant role (the most compelling evidence being from studies on identical twins raised apart). Some people are innately optimistic, while others see doom and gloom around every corner. Some love skydiving, while others would prefer to wear seat belts and safety helmets while resting on the couch. If the degree of expression of major personality traits such as optimism and risk taking has a major biological component, it follows that we would also experience different degrees of expression of involuntary mental sensations. Some people are seemingly predisposed to certainty—the "know-it-all"—while others are perpetual "prove-it-to-me" skeptics. A differing sense of agency can readily be seen in how we view voting. Some feel that a single vote can change world history; others feel that voting is the ultimate exercise in futility. Seen in this light, it is likely that how strongly each of us experiences a sense of causation will also vary widely from individual to individual.

To see how this variation in intensity of feelings of causation might play out, consider a recent study on the genetics of attention deficit hyperactivity disorder (ADHD). Scientists from Cardiff University found genetic differences between two groups of children—a normal control group and a group diagnosed with ADHD. According to the lead author, a professor of child and adolescent psychiatry, "Too often people dismiss ADHD as being down to bad parenting or poor diet. As a clinician it was clear to me this was unlikely to be the case. Now we can say with confidence that ADHD is a genetic disease and that the brains of children with this condition develop differently to the brains of other children."[3] The authors argue that the study proves that gene differences cause ADHD.

The actual data: fewer than one-fifth of 360 children with ADHD had a particular genetic variant, while more than four-fifths didn't. After reviewing the same data, others with equal background and expertise have come to an opposite conclusion—most ADHD must be caused by nongenetic factors.[4]

What fascinates me is that the study authors would feel so strongly about the causal relationship between genes and a complex, controversial, and ill-defined condition. Surely the authors must intellectually understand that behavior is a murky mixture of nature and nurture and rarely attributable to a single cause. It is easy to dismiss their interpretation as mistaking correlation with causation, but let me cautiously suggest an additional possibility. If each of us has his/her own innate ease or difficulty with which a sense of causation is triggered, the same data may generate different degrees of a sense of underlying causation in its readers. Though purely speculative, I have a strong suspicion that those with the most easily triggered innate sense of causation are more likely to reduce complex behavior to specific cause-and-effect relationships, while those with lesser degrees of an inherent sense of causation are more comfortable with ambiguous and paradoxical views of human nature. (Of course, for me to make any firm argument as to the cause of the authors' behavior would be to fall into the same trap.)

Unfortunately for science, there is no standard methodology for objectively studying subjective phenomena such as the mind. One investigator's possible correlation is another's absolute causation. The interpretation of the cause of subjective experience is the philosophical equivalent of asking every researcher if he/she sees the same red that you do. The degree and nature of neuroscientists' causal conclusions about the mind are as idiosyncratic as their experience of love, a sunset, or a piece of music.

There is a great irony that underlies modern neuroscience and philosophy: the stronger an individual's involuntary mental sense of self, agency, causation, and certainty, the greater that individual's belief that the mind can explain itself. Given what we understand about inherent biases and subliminal perceptual distortions, hiring the mind as a consultant for understanding the mind feels like the metaphoric equivalent of asking a known con man for his self-appraisal and letter of reference. In the end, we should start at the beginning, with the unpleasant but inescapable understanding that the less than perfectly reliable mind will always be both the mind's principal investigator and tool for investigation.

4 • Sensational Reason

It appears that, in single instances of the operation of bodies, we never can, by our utmost scrutiny, discover any thing but one event following another, without being able to comprehend any force or power by which the cause operates, or any connexion between it and its supposed effect. The same difficulty occurs in contemplating the operations of mind on body—where we observe the motion of the latter to follow upon the volition of the former, but are not able to observe or conceive the tie which binds together the motion and volition, or the energy by which the mind produces this effect. The authority of the will over its own faculties and ideas is not a whit more comprehensible.

—David Hume, An Enquiry Concerning the Human Understanding

I magine that you, an educated layperson with no particular area of expertise in climatology, are asked by a senior government official to help formulate public policy as it relates to climate change. Priding yourself on being rational and deliberate in your opinions, you tell the official that you'll "have to think about it."

What does "think about it" mean? As soon as the official's words reach your auditory processing regions, preliminary judgments begin jockeying for position. Long before (in brain time) you are fully aware of the meaning of the question, the question will activate those neural networks even remotely related to the subject of climate change. Your brain will sort through myriad bits of prior stored thoughts about the value of computer modeling and predictions, the integrity of scientists, the relevance of "climategate," feelings about "tree huggers," possible concerns for preservation of the polar bear and whether the photos of melting ice floes in the Arctic might be trumped up. It will factor in all your innate biases, your political leanings, your present mood, and your introspective beliefs about your personal character. Your brain will then deliver into consciousness the seemingly most appropriate initial image or thought along with whatever mental sensations this image triggers.[1] Before you can begin conscious deliberation, the playing field of your mind will already be littered with snap judgments and gut feelings.

To understand how this subliminal process works, I have borrowed a term, "hidden layer," from the artificial intelligence community. By mimicking the way the brain processes information, AI scientists have been able to build artificial neural networks (ANNs) that can convert speech to text, recognize faces, beat the best chess players, and win at *Jeopardy*. While standard computer programs work line by line, yes or no, all eventualities programmed in advance, the ANN takes an entirely different approach. The ANN is based upon a simple schematic: input→ hidden layer→ output. Interposed between incoming information and outputted signal is this metaphoric rather than distinct anatomic region that contains mathematical programs that learn from experience.[2] By weighting every input, the hidden layer creates decisions (outputs), which it then monitors for accuracy. This feedback from its outputs allows it to adjust its weighting of various components of a decision according to the success of the output.

For example, you're at a football game and your brain

receives an input from your retina that it believes might represent Sam's face. If it is indeed Sam, this positive feedback will trigger stronger and richer connections between various areas of the brain responsible for piecing together the recognition of his face. This enhanced "Hey, it's Sam" circuitry will make subsequent recognition of Sam easier and more accurate. But perhaps next time Sam has grown a mustache. The calculation weights all components—the distance between the eyes, the thickness of the eyebrows, the twinkle in the eye—then adds in an unknown variable: the new mustache. Your "Hey, it's Sam" circuitry fires up, but also recognizes that there's something different. Perhaps it isn't Sam, but just a look-alike. It needs confirmation to be sure. If it is Sam, feedback will incorporate the mustache into the "Hey, it's Sam" circuitry.

At the most basic physiological level, the hidden layer is the link between experience and learning, the primary conceptual mechanism for the generation and subsequent modification of all neural circuitry. But there's more to the hidden layer then the weighting of inputs from the outside world and sensory inputs from the body. The hidden layer isn't selective; it considers all inputs whatever their source. Conscious thoughts, psychological states, and memories are handled in the same way as more basic visual, auditory, or olfactory stimuli. All are thrown into a huge tangle of inputs that the hidden layer sorts out via its embedded computational skills.

Foremost among these inputs from the conscious mind are our various wishes, desires, and intentions, both past and present. If it's halftime at the football game and you're idly scanning the crowd, the hidden layer knows from values derived from past experience that you don't want to be bothered with irrelevant visual information. It has learned that you mainly want to be notified of familiar faces or threatening events. However, this time is different. You have a huge bet on the game and are solely interested in concentrating on whether or not to double your wager in the second half. You want to be left alone to make your calculations; you don't want to be notified of Sam's presence three rows down. The hidden layer takes this updated assignment and reweights incoming inputs. Sure enough, you fail to see Sam waving at you, then telling his wife that you've become too much of big shot to say hello. In effect, the conscious mind sends specific operating instructions to the brain; these commands become part of the hidden layer's revised job description.

Your conscious desires are but one component of your motivation; myriad unconscious factors will also be inputs guiding your hidden layer's behavior. What I wish to convey is not the primacy of the role of conscious intention, but that all conscious actions of the mind can be seen as inputs into a gigantic hidden layer that is already weighted with every aspect of our biology and prior experience.

Though we tend to think of perception as being exclusively involuntary, we also acknowledge the effect of attention on perception. Some of you may have seen the video asking you to count the number of times basketball players are bouncing a ball. In so doing, many of you will have failed to see the gorilla walking across the stage in front of the players. This phenomenon, referred to as "inattentional blindness,"[3] is commonly shown to demonstrate how attention affects perception. Attention is intention. "Count the dribbles" becomes the hidden layer's set of operating instructions. In carrying out your command, it fails to tell you about the gorilla.

On the other hand, if you are stumped by a particular problem, such as recalling a forgotten name or figuring out the solution to a novel or a troubled relationship, you will often find yourself coming up with an answer "out of the blue." Though this seems unintentional, like a gift from your muse or a spontaneous intuition, it isn't. Sometime earlier you had the conscious intention to resolve a problem but were unable to do so immediately. Your conscious intention was transported out of awareness and into your hidden layer, where it could work at its own pace, gathering together old and new inputs until it reached a possible solution. Only then did the answer appear in consciousness.

Let's return to the problem of what it means to "think about it." You've read the above paragraphs and glumly acknowledge that the starting point of your considerations begins out of sight in a simmering cognitive stew that isn't readily accessible to reverse engineering. No matter how much prodding and poking, you can't separate out the components into the original meat and potatoes. Nevertheless, you try your best to sweep away those unbidden feelings that you can recognize and begin your inquiry. You take a deep breathe and try to "clear your mind," a generally thankless task.

But what if you could create a clean slate and willfully start your deliberation free of unconscious biases? You would still run into yet other major constraints on conscious thought—limitations on working memory and processing speed.

The Fullness of Memory

Try remembering a phone number. On average, most of us can remember a seven-digit number, but longer numbers, such as for international calls, are more difficult. Irrespective of intelligence and education, our short-term (working) memory can hold only around seven bits of information at any instant. The most gifted have difficulty juggling more than nine or ten. To get around this limitation, we tend to organize material into meaningful groups or "chunks." When trying to keep a phone number in mind, we normally chunk the digits into three groups: the area code (three numbers), and then the local number, divided into one three-digit and one four-digit chunk. Three chunks are far easier to remember than a string of ten digits. On average, we can retain four chunks of information in short-term memory.

Imagine that your brain is equipped with an electronic clipboard equivalent to the random-access memory (RAM) of your computer. This clipboard can hold only a few chunks of information, yet it is your only tool for mentally carrying out conscious thought. If you want to add more chunks of information, you need to make room by transferring some data from the clipboard to your hard disk (long-term memory). To get a sense of this constraint, think of multiplying or dividing a string of numbers. For the easiest calculations, we merely rely on overlearned operations—grade-school multiplication tables. Once we extend beyond memorized computations and start manipulating numbers in our head, we quickly reach our mental limits. Looking for paper and pencil or calculator is the mental equivalent of being notified that RAM is full. The same memory constraints apply to our ability to manipulate symbols, words, and ideas.

Still worse for those who wish to retain their belief in complex conscious thought is the utterly depressing reality that short-term memory has a very short life expectancy. Within a minute or two, any information hovering in short-term memory either evaporates or is converted to longer-term memory. Once relegated to subconscious storage, the memory is subject to myriad subliminal influences. For any complex thought that takes more than a couple minutes, you have no way of knowing whether this thought has been subliminally affected by bias in its round-trip from conscious to unconscious and back again.

From the Age of Enlightenment onward, we have been told that man is rational. Only in the last century has the notion of unconscious cognition gained widespread acceptance. Even though we are increasingly aware of the myriad forms of perceptual and cognitive bias, the problem continues to be generally framed as "Here's another effect of unconscious brain activity on conscious thought." However, given the minuscule amount of available time and RAM, why not reframe the question as "What, if any, role does conscious thought play in our overall thinking?"

To see how this approach would develop, let's return to your contemplation of global climate change. On hearing the question, a handful of images pop into your mind and are pinned to your mental clipboard. You pick one—the plight of the polar bear—and start your deliberations. You come up with several reasons why the polar bear must be saved and a couple why it may have to go the way of the dodo bird. Your clipboard is now full; in order to toy with a second possibility— the speculative opportunity of purchasing future beachfront real estate in Greenland—you need to temporarily move the polar bear considerations into longer-term memory.

Say good-bye to your original deliberations. Unless you subscribe to the archaic and now widely disproven belief that the brain stores memories like MP3 files, you must grudgingly accept that memories undergo modification as they move in and out of long-term storage. While you are thinking about the beachfront property, the brief sight of a blue jay in your garden may trigger a set of new associations that will silently alter how you feel about the wildlife consequences of global warming. These new associations will be added to the neural circuitry containing your prior thoughts about the polar bear. This will be true even if you aren't consciously aware of or don't remember having seen the blue jay. After a few minutes of thought about Greenland, you decide against speculation and return to your thoughts about the polar bear. Your recall of your prior polar bear considerations now contains new or altered elements in the same way that a story morphs as it is passed around a campfire.

This necessary sequence of events undermines any reasonable likelihood that we are capable of carrying out purely concious, complex deliberations. We simply don't have the physiological firepower. Any thought that takes a couple minutes and/or involves more than a few items requires subconscious processing. Even if we knew the action of every neuron and synapse at every instant, we still wouldn't know what thoughts are actually carried out in consciousness as opposed to occurring subconsciously but feeling as though they take place in consciousness. Until such a hypothetical time when science uncovers the precise signature for conscious versus unconscious brain activity, any correlation between physical brain states and corresponding mental states is entirely dependent upon what subjects describe. Having a subject tell you that a thought is entirely consciously derived is no more valid than saying that thirst is a conscious decision.

In ruminating on the relationship between conscious and unconscious thought, I have gradually realized that the connecting link is our involuntary mental sensory system, and in particular those sensations of knowing, certainty, agency, choice, effort, and causation. Without these sensations, there would be no experience of a conscious thought. With them, we have a very strong sense that conscious thought is separate from those unconsciously generated thoughts that merely "pop into our heads," "are delivered by muses," or "suddenly occur to us." Consider how you feel when you are trying hard to solve a problem.

At the start of thought A you will experience the sense that you are thinking, the degree of effort involved, and that it is indeed your mind that is creating the thought. Without splitting semantic hairs, these are the mental states of ownership (my mind) creating a thought (sense of agency) by the act of paying attention and fully concentrating (sense of effort). Following this period in which you are only aware of your efforts, you may be rewarded with the appearance of a new thought B. This thought B will feel as if it is derived from thought A (feeling of causation) and will also be accompanied by a sense of the degree of rightness of this response (the feeling of knowing).

This sequence of mental sensations is the experience of a thought. We are not aware of the actual mechanics and steps involved; the nuts and bolts of cognition occur in utter silence in out-of-sight synapses and neuronal connections. If the act of thinking did have a built-in feeling tone, we would be aware of every subconscious cognitive action—an impossibly inefficient evolutionary design. Imagine sorting through all of our half-completed unconscious thoughts in order to avoid an oncoming diesel truck or saber-toothed tiger. Speed of response requires a lack of clutter.

Temporal Reorganization

A further complication: as our brain is quite skilled at reordering our sense of time, we cannot entirely trust the temporal sequence with which we experience our thoughts. When a baseball player sees the ball approach the plate and then swings at it, this perception of events has been dramatically altered. It is physiologically impossible to see the ball approach the plate and then initiate your swing successfully. Conscious perception of the approaching pitch and the time needed to react by swinging the bat take too long. To make the sequence of events feel meaningful, your brain reorders your perception; you consciously see the ball approach the plate before you swing despite the physiological reality that you began acting on the pitch and initiating your swing just a few milliseconds after the ball leaves the pitcher's hand.[4] This mechanism is applicable to all high-speed sports; it is estimated that the brain can subjectively alter the sequence of temporal experience by at least 120 milliseconds—certainly long enough to alter how we experience the sequence of our thoughts.[5]

Let's apply temporal reordering to the experienced sequence of events involved in a line of reasoning. You believe that your mind created the following sequence of events: first you thought about the polar bear (A), then about beachfront property in Greenland (B), but then this was formally dismissed in favor of further contemplation of the polar bear's future (C). From an experiential point of view, this sequence of thoughts proceeded in an orderly manner with each thought provoking the next. The experienced sequence of events: A → B → C. Yet we have no idea if this is how the sequence occurred at the level of brain operations.

The brain is capable of performing a number of operations simultaneously—the biological equivalent of parallel processing. Though we think of reasoning as one thought leading to another, this may not be the way that the brain actually works. Imagine that the hidden layer is a gigantic committee filled with individual influences that collectively generate a thought. Some represent biological predispositions, while others represent past experience. Everything from your DNA to your political leanings is given a vote. Incoming information is presented to the committee members, who cast their votes. The output—a new thought—would not necessarily require sequential processing. It may well be that all committee members voted simultaneously and that the tabulation occurred in an instant. If so, our conscious sense of a line of reasoning may not correspond to an actual brain event. In addition, if thought B is experienced as following A within a sufficiently short time span, we cannot exclude the possibility of temporal reorganization—that B actually happened first. As temporal reorganization happens in many high-speed activities such as sports, it may well occur with high-speed thought.

The Sequence of Thought

One of my professors during residency was a superb Scottish neurologist widely respected for his astute diagnostic skills yet notorious for his inability to outline the steps involved in arriving at a diagnosis. Some accused him of being mentally lazy because he refused to reduce his thinking to an algorithm or line of reasoning. When asked, he would shrug and with a slightly mocking expression say, "My mind doesn't work that way." Nevertheless, he remained the doctor's doctor. When their own health was on the line, physicians were willing to overlook his seemingly arbitrary decision-making process in favor of getting the best decision, however it was arrived at. I am not recommending shooting from the hip as the best way of thinking, but this neurologist's insistence that his reasoning didn't work sequentially has stuck with me. What if his seemingly offhand comment was a subtly presented insight—that decisions aren't necessarily linear, irrespective of how they feel?

It has taken me nearly a decade to realize that the fundamental impetus for believing in linear thought is the very strong sense that one thought causes another. Once again we are back to involuntary mental sensations. Combine the feeling of (1) a physical sense of self actively and deliberately creating its own thoughts, with (2) the feeling that these thoughts cause yet other thoughts, and you have a superb recipe for a powerful *sense of reason* taking place within a real mind. If we didn't have the feeling that our thoughts are causally connected, we would not have a working model of a mind, any more than having a TV monitor endless flash unconnected bits of random data would make for compelling viewing. Cause and effect is thought's driving narrative.

To create a functional mind, mental sensations must override any and all contradictory evidence. Dismissing these feelings as brain-generated conceits goes against our biological grain. In writing this paragraph, I have a strong sense of creating an argument point by point even as I realize that I may be expressing nothing more than an overall gestalt arising from an underlying simmering cognitive stew. In reality, there might not even be an underlying linear progression to my ideas. Perhaps this possibility of a complex thought occurring all at once as opposed to being a sequential line of reasoning explains the impetus behind describing a well-thought-out argument as "getting the larger picture."

To get a better handle on what reason might be, we need to have some understanding of the relationship between conscious and unconscious thought. Most important is to recognize that the distinction is purely arbitrary and unlikely to represent any fundamental biological differences. There is no decent evidence or reason to believe that the computational underpinnings of a thought are different depending on whether or not you are aware of them. To postulate that conscious thought is physically different from unconscious cognition, the mechanism for thinking would have to change as a thought moved in and out of conscious awareness. Other bodily processes don't function in this manner. A heart doesn't have a different contractile mechanism depending upon whether or not you are aware of it beating.

What are different are the degrees of computational ability and triggering mechanism(s). Though the brain has enormously greater computational capability than our conscious minds, it needs to be told what to do. Beyond reflexive actions and perhaps the simplest of calculations, the brain needs inputted operating instructions. A brain will not spontaneously try to plan a trip just because it can any more than it will slyly work on a novel or try to solve Fermat's last theorem just to surprise you. Its purpose is to accommodate its owner's wishes and desires, and have a rough idea of what might be a satisfactory initial response. When your brain receives word that you have next week off, it springs into action. It knows what a week off "means" to you and what might constitute a good set of possibilities to show you. Like a good salesman, it will sort through your past memories and preferences to suggest those items most likely to be pleasing to you.

To see this interaction in computational terms, consider how you perform a Google search. No matter how powerful the Google search engine, it needs you to type in some instructions to prompt it into action. A blank search box yields nothing. Once you've entered a key phrase or two, the search engine has been supplied with a goal. It knows what you want; you can sit back and let the search engine perform the necessary calculations. Complex search box entries aren't necessary; in fact, they may be counterproductive. A few key words are enough. Conscious thought is the equivalent of typing a few directions into your brain's search box. This action fits nicely with its available processing power; a few chunks of memory entered into the search box are enough to trigger the most complex computations imaginable. Ideas move in and out of consciousness, get stored, modified, retrieved, consciously reworked, sent back for temporary storage, further modified, recalled, and so on, until a final conclusion is reached. It is through multiple iterations of this process that we end up with what we feel is a consciously derived line of reasoning.

Though no metaphor is entirely adequate, let me offer the hard-to-quantify relationship between a symphony conductor and the members of the orchestra. Let the conductor stand for conscious cognition, musicians for the committee members of the hidden layer, with music representing the final thought. A conductor/musical director picks out which musical selections the orchestra will perform and chooses the tempo, coloration, and desired interpretation for each piece. Each orchestra member brings to the performance his own prior knowledge of the work and his playing skills, personal beliefs, and aesthetics, including his own preferences for the optimal performance. Nevertheless, the collection of musicians looks to the conductor for additional input and ongoing guidance. In listening to the performance, it is impossible to arbitrarily determine the relative roles of the conductor and the musicians.

There is no moment in the performance when the conductor is either making or capable of making the actual music (other than humming a bar here and there). Though only one input among many into how the musicians will play, the conductor can be tremendously influential. Musicians will try to follow his lead even though they will also be guided by their own habits and preferences. The quality of the performance and the degree to which it follows the conductor's guidance will vary from moment to moment, player to player, orchestra to orchestra. No matter how great a conductor's influence over the players, no absolute predictions can be made in advance. The resulting performance can be expected to deviate from his intentions as orchestra members respond positively or negatively to their surroundings, to the interaction with other players, to audience enthusiasm or the latest news on salary negotiations. From the point of view of the audience, the music is being made by both the conductor and the orchestra. They are acting as a unit. We watch the conductor's baton descend and his arms flap about, and, guided by our own individual sense of causation, we have a strong sense that these movements are causing the orchestra's particular interpretation.

We are conductors of our unconscious cognition. We provide intention, direction, and sets of instructions that get the unconscious cognitive ball rolling and then provide ongoing guidance. The brain works according to its own innate and acquired methods, but takes its cues from conscious input. Without the conductor, the orchestra wouldn't know what to play. Without conscious input, the brain wouldn't know what problem to address. Without the orchestra, there would be no music. Without unconscious cognition, there would be no complex thought. In short, the experience of strictly conscious complex thought is pure illusion—a sensory misperception generated by a host of involuntary mental sensations.

5 • Logic's Reason

Ignorance more frequently begets confidence than does knowledge.

—Charles Darwin, The Descent of Man

Other than in the ivory tower of academic philosophy, few if any of us feel a need to challenge the assumption that logic, like mathematical proofs, has at its roots a purity not affected by subliminal biases and perceptual influences. But if logic is the formal analysis of reasoning, and we experience reasoning via involuntary mental sensations, what does this tell us about the underlying nature of logic?

After recently watching a few online introductory philosophy courses, my first reaction was that a fair number of age-old philosophical arguments seemed nonsensical. If you have an infinite number, what is infinity plus one? If you have a heap of sand and take away the sand grain by grain, when do you no longer have a heap? What came before the beginning? But the one I continue to have the most trouble wrapping my practical mind around is the question central to philosophical, theological, and scientific inquiry: How does something arise from nothing? Whether thinking about the origin of the universe or about the possibility of a Prime Mover or Creator, this question feels chock-full of meaning—from an answer to our origins to a possible clue as to a "purpose" for our lives. But is this really a meaningful question? How would we know? The form and syntax of the question are straightforward, without unnecessary jargon or ambiguity. World-class philosophers skilled in the rigors of logic find the question suitable for a lifetime of pursuit. But just because a question seems to make sense doesn't mean that it represents a real-life problem or is even logical. Consider Wittgenstein's famous quip, "It is afternoon on the sun."

Fuzzy Logic

The best example of our relative inability to assess the logic of our thoughts can be seen in a 1999 study by Cornell psychologists Justin Kruger and David Dunning, "Unskilled and Unaware of It: How Difficulties in Recognizing One's Own Incompetence Lead to Inflated Self-Assessments."[1]

The researchers asked a group of Cornell undergraduates to take a logical reasoning self-assessment test based upon a twenty-item set of questions from the Law School Admissions Test (LSAT). At the completion of the test but before receiving the results, students were asked to compare their general logical reasoning ability to that of their fellow classmates. They also were asked to guess how many test questions they had answered correctly. On average, participants placed themselves in the sixty-sixth percentile, revealing that most of us tend to overestimate our skills somewhat (the so-called above-average effect). Those in the bottom 25 percent consistently overestimated their ability to the greatest degree.[2] Individuals who scored at or below the twelfth percentile believed that their general reasoning abilities fell at the sixty-eighth percentile. In addition, those in the bottom quartile overestimated the number of test items they got right by nearly 50 percent. Conversely, the top-quartile participants overestimated the ability of the lower-quartile students. "Because top quartile participants performed so adeptly, they assume the same is true of their peers," said the study authors.

In a second set of studies students were given a packet of five sets of ungraded test answers by students from the earlier study. They were told that these test packets reflected the full range of performances that their peers had achieved. All participants were then reshown their own tests and asked to re-rate their ability and performance. In effect, they were shown the total spectrum of possible answers and performance—evidence that answers different from their own might very well belong to the top performers—and then were given the opportunity to reassess their own performances. Despite this new knowledge, those in the bottom quartile did not change their own self-estimates. Knowing that almost four-fifths of the ungraded test results were different from their own, those in the bottom quartile still believed that they had performed better than more than two-thirds of the test takers.

Dunning and Kruger's study quantified the dual burden of incompetency in logical analysis. The lack of skills also prevents the individual from recognizing this incompetence. The authors' conclusion: "People who lack the knowledge or wisdom to perform well are often unaware of this fact. That is, the same incompetence that leads them to make wrong choices also deprives them of the savvy necessary to recognize competence, be it their own or anyone else's."

This "Dunning-Kruger" effect raises a serious cognitive dilemma. Those who are most impaired in the use of logic are the most likely to overestimate their logical abilities and underestimate the logical skills of others. (Does this sound like much of what passes for public discourse?) Unless we can figure out a way to compassionately identify those so affected and improve their reasoning skills, we cannot expect them to recognize when their reasoning has gone awry. At the same time, there's no point in name-calling; after all, none of us can be certain that we are not one of the affected. Further, we need to recognize that these difficulties with logical thinking don't necessarily reflect an overall stupidity. The students in this study had been admitted to a prestigious Ivy League university, presumably on the basis of having done well in high school and on college boards and aptitude tests.

It is easy to jump to a wide variety of psychological explanations for the "Dunning-Kruger" effect. But what if there is a more direct physiological relationship between illogic and excessive self-confidence in one's logic?

Take a moment to think how we arrive at a logical decision. With a complex or convoluted question, it takes time and trial and error to sort out logical fallacies that may not be immediately obvious. The more difficult the problem, the better off we are looking at it from as many perspectives as possible. A good imagination, an open mind, and a willingness to avoid hasty conclusions are essential. But what if you experience a premature sense of having the right answer? Or an answer that "feels" better, more elegant, or more familiar than another? Once these feelings are firmly in place, you are far less likely to suspect that a line of reasoning might be logically flawed. We've all taken a multiple-choice test in which one answer seems more likely to be right simply because it feels more familiar than the others. Familiarity is the brain's way of telling you that the answer bears some resemblance to a previously stored memory or bit of data.

Critical thinking is a skill acquired in the same way as learning how to play the piano. Just as we establish circuitry for piano playing, we develop representational maps for how to think. If a present line of reasoning matches how we have thought in the past, it is more likely to generate a sense of familiarity and correctness. Conversely, trying a new line of reasoning is likely to feel strange, unfamiliar, and incorrect. The greater our reliance on such involuntary mental sensations as familiarity and sense of correctness, the more likely we are to persist with the belief that our logic is impeccable even when presented with potentially contradictory evidence (as in the Dunning-Kruger study).

Beauty in Numbers

There are yet other mental sensations affecting logic that rarely make the headlines. It has long been believed that mathematicians and scientists use beauty as a clue as to whether or not a judgment is true. Some mathematicians have suggested that a sense of beauty is the primary motivation for mathematical discoveries.[3] World-class logician Bertrand Russell once wrote: "Mathematics, rightly viewed, possesses not only truth, but supreme beauty—a beauty cold and austere, like that of sculpture, without appeal to any part of our weaker nature, without the gorgeous trappings of painting or music, yet sublimely pure, and capable of a stern perfection such as only the greatest art can show. The true spirit of delight, the exaltation, the sense of being more than Man, which is the touchstone of the highest excellence, is to be found in mathematics as surely as poetry."[4] Paul Erdos, famous mathematician, has said, "Why are numbers beautiful? It's like asking why is Beethoven's Ninth Symphony beautiful? If you don't see why, someone can't tell you. I know numbers are beautiful. If they aren't beautiful, nothing is."[5] This feeling of beauty is qualitatively different from feelings of certainty. As Erdos points out, there is a sense of beauty in numbers unrelated to any particular meaning or conclusion. The numbers themselves are beautiful. Russell captures this purely aesthetic sense in describing a pleasing equation as being comparable to great art.

To test this hypothetical relationship between beauty and perceived truth, in 2004, researchers headed by Rolf Reber from the University of Bergen, Norway, studied the effects of symmetry on subjects' perception of correctness of simple calculations. Their choice to study symmetry as a stand-in for beauty was based on the general preference for symmetry in humans, other primates, and a wide variety of other species including bumblebees, fish, and birds, as well as the oft-noted relationship between perceived symmetry and "truth in mathematics." (Perhaps it is a leap to correlate preference for symmetry with beauty, but I appreciate the elegance of the study, so I have accepted the premise. If I'm wrong, use my misunderstanding as further evidence for my argument.)

The researchers constructed visual representations of addition problems by using clusters of dots. For example, a subject would be shown 10 dots + 20 dots and then the answer, written as 30 dots. Half of the additions shown were correct; the others were wrong, such as 12 dots + 21 dots equaled 27 dots. Half of the additions were created using symmetric dot patterns and the other half with asymmetric patterns. Each set of dots was presented for less than two seconds—not enough time to count the dots. Immediately after the image disappeared, subjects were asked whether or not the addition was correct. Subjects consistently felt that a symmetrical presentation was more likely to be correct than an asymmetrical presentation of the same dots. Reber believes that symmetry allows for a faster speed of neural processing, which in turn contributes to the perceived accuracy of a statement.[6] In other words, faster neural processing increased the likelihood that the subjects would interpret an answer as correct.

To put this idea to the test, his team devised a simple experiment. On a video monitor they briefly presented a word followed by a possible anagram of that word. Subjects were asked to score the likelihood that the second word was indeed an appropriate anagram of the first. Sure enough, the quicker the appearance of the second word, the more likely it was felt to be a correct response (a suitable anagram), irrespective of whether or not the answer actually was a correct anagram. Answers flashed at 50 milliseconds were far more likely to be judged correct than those shown at 150 milliseconds. A similar correlation was found when presenting math equations followed by possible answers. A 50-millisecond lag in presentation was enough to significantly reduce the likelihood that an answer would be judged as correct.

The sense of beauty and truth as features of processing speed helps explain our tendencies to judge the familiar as correct, why deeply ingrained patterns of thinking are likely to feel more correct than alternative possibilities, why we stick with brand names, and why new information and ideas are less likely to feel correct than old ideas with which we are comfortable.[7] A physiological link between a sense of familiarity, habit, and aesthetic preference is the faster speed with which well-oiled neural circuitry can process previously mulled-over ideas in comparison with the longer time frame necessary to assimilate and process a brand-new idea. It is as though the relationship between processing speed, experience, and sense of truth can be summed up in the cliche "tried and true."

A second feature of the study—the decay time for the sense of correctness—is also worth noting. Information presented at 150 milliseconds was far less likely to trigger a sense of being correct than information presented at 50 milliseconds. To put this into everyday experience, conscious perception takes several hundred milliseconds. Involuntary mental sensations occur much sooner. You sense a looming threat before you consciously perceive an out-of-control truck about to run into your car. A name feels familiar before you have fully comprehended the name. Given this lag time, first impressions and instantaneous gut feelings have a built-in advantage over slower opinions derived from conscious thought.

If quick decisions are those most likely to save a life in an emergency, and immediate action is predicated upon feeling you are making the right decision, the association between speed and correctness would have real evolutionary value. Unfortunately, this evolutionary advantage works best with simple problems, such as deciding to duck to avoid an incoming spear, but is less well-suited for addressing complex modern issues such as climate change and the risks of nuclear power. It's no surprise that shapers of public sentiment avoid complexity and subtlety, and confine their presentations to the briefest of sound bites—an approach consistent with Reber's observation that the quicker the presentation, the more likely it is to feel correct. Whether with Machiavellian intent or because they possess unsuspected neurophysiological savvy, opinion makers have learned that nuance and subtlety are the enemy of the "aha."

At the risk of wearing out the brain-as-computer metaphor, most of us implicitly suspect that intelligence equates with a faster rate of processing information. Smartness means that you catch on quicker, grasp a new idea more facilely, get it right away, etc. But if Reber's studies are right, intelligence is a dual-edged sword. If we lack basic cognitive skills, we are more likely, as in the bottom quartile in the Dunning-Kruger study, to overestimate our logical skills. And even if we are very smart and have a superfast brain, the quicker the arrival of a potential answer to a problem, the more likely we will be to overestimate its correctness. The implication is stunning: the smartest among us also have an inbuilt mechanism favoring overestimation of our reasoning skills and the correctness of our thoughts.

Optics of the Mind

Suppose that you're a renowned physicist who has devoted his life's work to unraveling the mysteries of the cosmos. You have made some first-class discoveries in the field, and are spurred on, despite overwhelming physical problems, to make some final sense out of your work. How might the very elegance and beauty of your proposed "theory of everything" affect your judgment of the logic and truth of your ideas? To see the interrelationship between involuntary mental sensations and the belief in one's impeccable reason and logic, let's take a look at some of the claims of one of the preeminent thinkers of our time, Stephen Hawking. In his recent book, *The Grand Design,* Hawking makes the rather remarkable claim that "because there is a law such as gravity, the universe can and will create itself from nothing. Spontaneous creation is the reason there is something rather than nothing, why the universe exists, why we exist."[8]

Take a moment to consider why you would consider the concept "something from nothing" to be a meaningful proposition rather than a problem of semantics, or worse, utter nonsense, the modern-day equivalent of alchemy. At first glance, it makes no sense to talk of spontaneous creation arising from nothing without undoing basic laws of physics.[9] But for the moment leave aside the notion of what nothing "is" (including quantum definitions of nothing), and simply consider how such a question originates in the brain. To begin, try to envision the big bang. In all likelihood you conjure up an image of a dense ball of matter outlined against a different-colored background. Our visual cortex—which creates the mind's eye—projects the shape of an object shape by creating borders and edges against a contrasting background. Without a background, we are unable to have a foreground image.

Most of us will come up with a sense of background grayness or darkness, but it's impossible to generate an image of nothingness. All mental images have some shape in the same way that we cannot see a color in the absence of some form. The best we can manage is a sense of empty space—which is itself a form without content. Any starting point for visually conceptualizing the moment of the birth of our universe runs up against the physical limits of the mind's eye's need for a background against which a foreground can be visualized.

To see how all-pervasive this difficulty mentally visualizing "nothing" can be, think of a simple subtraction problem. If you have four apples and take four apples away, you have no apples, but you still have the empty space that once housed the apples. If our mind's eye envisions an object and then we take the object away, we are left with the remaining space serving as the visual brackets (background) against which we saw the object. The same goes for calculations. Imagine an equation $4 - 4 =$ and the answer is left blank. As this makes no sense, we have constructed the concept of zero. Zero is not nothing. A vacuum is not nothing; it is an empty space devoid of any substance. At an experiential level, we will always be left with any representation of nothingness occupying time and space.

For me, the question of how something arises from nothing is a philosophical and intellectual dead end. The failure of 2,500 years of inquiry should be a warning that the question isn't resolvable, has some inherent logical flaw, is a problem of semantics,[10] or is a fundamental paradox arising out of the way that our visual cortex creates our "mind's eye" worldview.

(Perhaps there are those with extraordinary imaginative powers who can think exclusively at the abstract level, but for the vast majority of us, a worldview requires some sort of image— hence the term "worldview.") Of course, I cannot test this hypothesis in any meaningful manner; I am constrained by the very visual imagery that prevents alternative viewpoints. At the same time, my brain will prompt me to keep asking the question, as it is hardwired to seek resolution of ambiguous visual representations.[11] I suspect that this biological urge to explain the mind's eye background is a major factor perpetuating the seemingly most pressing philosophical/theological issue that confronts modern man.

Now watch what happens if you try to step away from this biologically generated foreground/background mental imagery. To avoid the problem of what lies outside the edges of the universe, Hawking proposes that the universe has no boundaries. To do this he would ask us to accept a different view of the universe. His visual metaphor: the universe has a shape that, like the earth, can be continuously circumnavigated without running into a wall, a cliff, or a chasm. If it can be navigated in perpetuity, it is effectively without a surface boundary. Though this imagery seems to satisfy Hawking, I am unable to grasp how the lack of boundaries along the "surface" of the universe says anything about what is beyond or before the universe.

In the opening chapter I described how the first-person point of view was generated by the brain's mental sensory system. Here we can see how a difference in perspective becomes the starting point for a theory designed to solve a problem that itself may be nothing more than a mind's eye–generated question.[12] But even though Hawking has spent most of his career working out the details of his theory, his alternative mind's eye perspective can't overcome the more compelling universal mind's eye foreground/background problem. In the end, Hawking is forced to wrestle with what to say about the background surrounding his no-boundary universe. In trying to make his alternative perspective explain our default perspective, he has painted himself into the logical corner of insisting on spontaneous creation and that gravity somehow existed prior to time and space.[13]

I am not in a position to challenge Hawking's theoretical premises; I have no real understanding of math or physics. My point isn't about the truth of his theory. What concerns me is the extent to which ignorance (or intentional disregard) for how the brain creates thought is morphed into a deification of pure reason. According to a recent press release, Hawking has said that his explanation of the laws governing us and our universe is currently the only viable candidate for a complete "theory of everything." If confirmed, it will be the unified theory that Einstein was looking for, seen as the ultimate triumph of human reason.[14]

I have no desire to single out Stephen Hawking for criticism. (I am a great admirer of his work and of his heroic spirit in living with one of the most incapacitating and demoralizing of neurological diseases.) My interest is in underscoring how an operational conception of the mind is the beginning point for all theories—whether talking about the cosmos, climate change, or the nature of consciousness. Theories should not begin with assumptions about the subject under inquiry; they should begin with a close look at the tool—the mind—that creates these assumptions. Otherwise it is a short step to believing in the spontaneous creation of something from nothing.

6 • Metacognition

Besides, who is to say that the feelings he writes in his diary are his true feelings? Who is to say that at each moment while the pen moves he is truly himself? At one moment he might truly be himself, at another he might simply be making things up. How can he know for sure? Why should he even want to know for sure?

—J. M. Coetzee, Youth

In recent years, rationality has taken some major hits, ranging from the revelation of myriad inherent biases, to the disruptive role emotions can play in so-called rational decision making. Self-improvement movements such as the practice of emotional intelligence increasingly focus on trying to control our wayward emotions. We can learn to count to ten, take a time-out, deep breathe, and avoid making major decisions when we are feeling vengeful or short-tempered. Ideally, we should be able to develop similar techniques for recognizing and coping with potentially unreliable mental sensations.

However, there's a fundamental difference between emotions and mental sensations. We cannot step back from mental sensations; they are the very means with which we judge our mental states. In *On Being Certain*, I emphasized the underlying biology that makes overcoming an unwarranted sense of conviction so difficult. Unfortunately, this is equally true for other mental sensations. Consider how personal perception of causation (or lack of it) is central to our opinion on global climate change.

Most climate scientists believe that "ferocious storms are consistent with forecasts that a heating planet will produce more frequent and more intense weather events."[1] Nevertheless, for some, the severe snowstorm in the Northeast in the winter of 2009 was evidence against global warming. According to Senator James M. Inhofe, "If it's snowing more, it's colder outside, so global warming isn't occurring."[2] Donald Trump, referring to the severe snowstorm, felt that the Nobel committee should strip Al Gore of his Nobel Peace Prize.[3] Senator Jeff Bingaman, chairman of the Energy and Natural Resources Committee, admitted that the heavy snowfall made it more difficult to argue that global warming is an imminent danger.

Eyewitness accounts are notoriously unreliable, subject to all the idiosyncrasies of individual perception. Determinations of causation not subjected to harsh scientific scrutiny are no different. In effect, perception-based determinations of causation are eyewitness accounts. It is easy to dismiss such global warming denials as political posturing, but this would be shortsighted. Rather than viewing contrary opinions as arising solely from devious motivation, gross stupidity, or a host of psychological maladies, we might also consider their physiological underpinnings. Combine the mental sensations of strong feelings of certainty and causation with basic biological tendencies to avoid ambiguity and reduce complex issues to manageable "chunks" (either of information or opinions), and then toss in a generous helping of personality traits (inflexibility, lack of empathy and concern for future generations, pride of ownership in an idea), and you have a first-class biologically mediated recipe for climate denial. Snow in the street becomes prima facie evidence that things aren't heating up. (The tragic circularity of political discourse: what starts out as a political agenda ends up as a loud voice in our hidden layers, and conversely, what starts out as a misunderstanding of causation ends up as political agenda.)

Just as mental sensations affect our thoughts, how we think about mental sensations affects our understanding of the mind. Stop me if you've never questioned whether your pet turtle or iguana misses you when you've been away. Or watched two geese closely guard their young and wondered if they have any sense of altruism, compassion, or self-sacrifice, or realize that they're monogamous (so many geese, so little time, and yet they bond for life). Whether explaining to your young child about her robotic dog's feelings, or trying to decide if a clinically unconscious family member might have undetected awareness, we invariably end up either assigning or denying others agency, intention, and sense of self depending upon our understanding of what such mental states "are."

Do you feel that a plant leaning toward the sun is doing so out of intention and sense of purpose? If not, why not? Many of us would start with the most basic assumption—the plant has no central nervous system, so it can't have intentions. The leaning is entirely reflexive—a straightforward demonstration of phototropism. But if an elderly hard-of-hearing aunt leans toward you to hear you better, you believe that her behavior is intentional. Even if you suspect that it's entirely automatic, you're still likely to assign to it some purpose and intention, albeit at an unconscious level.

If IBM's supercomputer beats Gary Kasparov at chess, we don't believe that the computer has any understanding of what it was doing, or any sense of intention or agency. It is just carrying out the intentions of the programmer. Now imagine interviewing a previously unknown player who's handily beaten Kasparov. You ask him how he managed to beat one of the best players in history. The man shrugs and says, "I have studied and memorized all of the chess games ever played. I then made my moves according to a complex calculation of probabilities of success. I don't understand a thing about chess." Would you believe that the man understood nothing about chess, or would you suspect he's being disingenuous?

Whether developing academic theories of mind or considering the role of fMRI in mind reading, we are in the peculiar position of using our own experience and understanding of involuntary mental sensations to conceptualize how these states may or may not be present in others. To get a flavor for real-world implications, consider how our conception of an animal's mind affects how we treat it.

Sensing Our Uniqueness

Though an alleged dog lover, Descartes wrote that dogs were totally lacking in consciousness and described his own dog as an automaton.[4] Once you deny animal consciousness, it is a short step to believing that animals don't experience pain and suffering. It is hard to understand how anyone with a set of eyes or a trace of empathy could come to this conclusion, yet an integral part of my med school training was doing physiology experiments on often poorly anesthetized dogs. Perhaps my worst memory is of removing the pancreas of a "donated" stray from the animal shelter so that we could witness firsthand her progressive deterioration and ultimate death from surgically induced diabetes. I can still see her cowering in her cage, whimpering. I cringe at the look of betrayal in her eyes.

What animals feel and think has been the subject of much controversy since the days of Aristotle. That we still don't have a consensus opinion is understandable. We can judge animals only by observing their behavior, as they cannot describe what they are experiencing, and observing behavior can never lead to a consensus opinion any more than we can expect unanimity of opinion in eyewitness accounts.

In the introduction to his 2008 book *Human: The Science Behind What Makes Us Unique*, Michael Gazzaniga, an acknowledged pioneer in brain research and director of the University of California, Santa Barbara, Sage Center for the Study of the Mind,[5] writes, "Let us start the journey of understanding why humans are special. . . . Though we are made of the same chemicals, with the same physiological reactions, we are very different from other animals." His central assertion is that we have undergone a physical transformation, the equivalent of a phase shift that makes ice and mist chemically similar but different "in their reality and form." As evidence for our uniqueness, he cites the nature of "our brains, our minds, our social world, our feelings, our artistic endeavors, our capacity to infer agency, our consciousness."[6]

Most of us have a strong belief in the basic tenets of evolutionary biology. We generally accept that we evolved from other animals and are not the only creatures with eyes, ears, or pain fibers. We are not the only animals that appreciate symmetry, demonstrate artistry, and have well-developed social skills. And yet, buoyed up by a deeply felt sense of a unique self, first-rate scientists like Gazzaniga feel compelled to proclaim our essential difference from the rest of the animal kingdom. Perhaps his use of "unique" in his book's subtitle is nothing more than a reflection of the publisher's marketing efforts, but, to me, "unique" has the faint whiff of self-congratulation and "speciesism." Worse, it is this same sense of uniqueness that drives the most extreme arguments of creationists. Listen to Sarah Palin: "I do not believe in the theory that human beings—thinking, loving beings—originated from fish that sprouted legs and crawled out of the sea" or from "monkeys who eventually swung down from the trees."[7]

The rebuttal to creationism isn't more scientific evidence. How much wiser it would be for neuroscientists to emphasize that feelings of uniqueness are as illusory as other involuntary mental sensations. As long as our preeminent students of the mind feel obliged to tell us about the scientific reasons for our uniqueness, they are conspiring with the enemy, those most dedicated to the advancement of antiscientific sentiment.

To see the inherent difficulties in knowing the mind of an animal, take a look at the male bowerbird's complex courting behavior. Best observed in the rain forests of New Guinea and eastern Australia, the pigeon-sized bird forgoes gaudy plumage or snappy birdsong to attract the ladies; instead, he constructs an elaborate forest floor structure—a bower—from moss, twigs, and leaves. He then decorates the bower with colorful odds and ends from feathers and pebbles to berries and shells.[8] The collective effect is often dazzling, and has been compared to a well-decorated bachelor pad. How we interpret this behavior depends on whether or not we choose to believe that the bowerbird is exhibiting intentionality, appreciation of aesthetics, and the desire for artistic expression or is merely acting according to hardwired reflexes. Instinct or cunning is a function of the observer's mind, not a scientific fact.

Few would seriously consider that an amoeba that withdraws from a noxious stimulus has consciously calculated the best path of retreat. We are likely to have a greater degree of disagreement when we see a lobster flailing about when thrown into a pot of boiling water. As we move up the evolutionary ladder, such judgments become more difficult. Inherent in these assessments is some notion of greater and lesser—we are more than an amoeba, but how about dolphins? Whales? Parrots? For the most part, our judgment of primitive versus advanced is based on how closely an animal exhibits human behavior as opposed to how closely we exhibit animal behavior. Maybe a bowerbird isn't a Picasso, but if the goal is attracting females with a particular aesthetic sensibility, the bowerbird should be considered an artistic success. Perhaps we'd judge the bowerbird differently if it wore a beret.

In his 1990 book *Animal Liberation*, animal rights activist and philosopher Peter Singer wrote:

> Do animals other than humans feel pain? How do we know? Well, how do we know if anyone, human or nonhuman, feels pain? We know that we ourselves can feel pain. We know this from the direct experience of pain that we have when, for instance, somebody presses a lighted cigarette against the back of our hand. But how do we know that anyone else feels pain? We cannot directly experience anyone else's pain, whether that "anyone" is our best friend or a stray dog. Pain is a state of consciousness, a "mental event," and as such it can never be observed. Behavior like writhing, screaming, or drawing one's hand away from the lighted cigarette is not pain itself; nor are the recordings a neurologist might make of activity within the brain observations of pain itself. Pain is something that we feel, and we can only infer that others are feeling it from various external indications."[9]

If behavioral observations—such as assessing an animal's pain—aren't sufficiently reliable, can science offer a more reasonable approach? New Zealand neurophysiologist Craig Johnson thinks so. In 2009, he reported that "brain signals have shown that calves do appear to feel pain when slaughtered."[10] Johnson's award-winning work was predicated upon the use of brain wave testing (EEG). Previous studies involving human volunteers and several other mammalian species have shown certain EEG patterns when the subjects receive painful stimuli. In order to spare the animals any discomfort, Johnson anesthetized the calves before slitting their throats. As he had postulated, these animals demonstrated typical "electrical pain signals" when their throats were slit. His conclusion: the calves would have experienced pain if they had been awake.

Based upon an EEG finding that correlates with clinical pain descriptions from awake human volunteers, Johnson has presumed that the EEG pattern represents the mental state of pain. But it makes no sense to refer to a brain wave pattern in an anesthetized animal as representing the unconscious manifestation of pain. Pain is a conscious experience. Imagine two people getting an identical stubbed toe. One has just lost his job and his wife's left him; he howls in agony. The other has just received word that he's won the megalottery; he doesn't even notice the stubbed toe. The inputs are identical; the neurological precursors of pain will be identical. But the mental experiences will be quite different. This study is a perfect example of confusing the brain with the mind. Knowing what is occurring at the level of neuronal function cannot tell us what the animal is experiencing.

One of the tragic sidebars of this study is the researcher's admirable attempt to use an EEG pattern to convince religious leaders that animals could experience pain. According to Johnson, the results of the study weren't a surprise, "but in terms of the religious community, they are adamant animals don't experience any pain, so the results might be a surprise to them." As rebuttal, a spokesperson for a religious slaughterhouse cited a prior study from the University of Hannover, Germany, that concluded, based on EEG patterns, that one slaughtering technique was more humane than another.[11] It is a remarkable leap of faith and a profoundly reductionist view of the inner life of an animal to believe that a decision governing the humane killing of animals could be determined by interpreting brain wave patterns.

This conflict between the nonobjectivity of behavioral observations and the inability of science to bridge the mind-brain gap doesn't lend itself to a coherent solution. Neuroscientists can propose yet further physiological investigations while philosophers can offer up a never-ending stream of thought experiments, but there is no airtight resolution to the problem of subjectivity trying to objectify itself. As we will see shortly, this notion of neural correlates of conscious mental states is at the heart of a number of neuroscience misconceptions ranging from assessments of consciousness, to assigning causation to certain pain syndromes, to the claims that morality can be ascertained scientifically.

Chipping Away at Intelligence

If the history of downplaying the mental capabilities of animals is a lesson in how not to think about other creatures, let's look at how far this lesson should extend. The logical extreme would be to consider machine intelligence—the inevitable next step in conceptualizing the extent of what a mind can be. To do so, I've chosen to look at one of the more intriguing concerns of modern man: our relationship to computers. HAL (the robot in the movie *2001: A Space Odyssey*) has become an icon for the evolving relationship between man and machine. Some super-optimists are convinced that machine intelligence will eventually far surpass our own. For others, the possibility of truly intelligent robots is a direct threat to man's most unique quality—his mind.

To see how we differ from computers, each of us needs to decide what, if anything, a computer actually understands.

The first step would be to agree upon the operational definition of understanding. But any cognitive process is composed of two components: the actual computations and the sensory experience—the sense of understanding—that arises from these computations. This relationship isn't as straightforward as one might hope. Correct computations may not be associated with a sense of understanding; we've all used mental effort to fill out a complicated tax return or reconfigure and reboot a malfunctioning computer modem without having any sense that we understand what we are doing. Incorrect computations are commonly associated with an unwarranted sense of understanding, as seen in the Dunning-Kruger effect. Understanding and intelligent thought are not synonymous; they represent distinctly different concepts and mechanisms. Even if your brain has figured out the most spectacularly brilliant solution to an age-old problem, at the neuronal level, it is still just a computation. The associated involuntary feeling of understanding is how you experience your brain's calculations. It is this understanding that varies from an everyday "aha" moment to the once-in-a-lifetime epiphany.

In 1980, philosopher John Searle published a controversial thought experiment—the Chinese room argument—in order to explain why artificial intelligence does not represent understanding. Searle imagined a strictly English-speaking man inside a locked room. The man is given a set of instructional manuals telling him exactly how to process incoming information written in Chinese; this response is then outputted in Chinese. As he does not understand Chinese, his responses are based solely on following instructions from the manual (written in English); he understands neither the input nor the output, but he is able to perform the task appropriately. The outputted responses will make perfect sense to a reader of Chinese, but will have no meaning for the man in the locked room. Searle's argument is that no matter how accurately a person can follow instructions and generate seemingly meaningful responses, if he does not understand the semantics and meaning of his responses, he cannot be said to have understood anything. Over the ensuing thirty-plus years philosophers have generated mountains of arguments both for and against Searle's position.[12]

The entire puzzle of whether or not computers understand would disappear if you accepted that understanding is a mental experience dependent on a mental sensory system. As computers lack such human sensory systems, no one should expect understanding to be present. Asking the question is meaningless. At the same time, the absence of computer understanding does not address the question of whether or not machine computations can be considered a form of intelligent thought. Most of us would agree that getting your malfunctioning modem up and running without having the slightest clue of how a modem's settings work is an example of intelligent problem solving seldom if ever associated with an aha or a deep sense of real understanding.

Computers have their own kind of intelligence, based upon whatever data and feedback that they've accumulated. We take it for granted that they are capable of calculations far beyond what humans can do and that these computational skills will continue to improve at a dramatic, perhaps even a geometric, rate. We accept without much discussion that computers can beat the best *Jeopardy*, backgammon, and chess players and generate (via modeling) dramatic insights into such diverse topics as game theory, climate change, and cosmology.[13] The difference lies in the nature of these calculations. Without sentience and associated mental sensory systems, the most technologically advanced computer equipped with the most sophisticated AI program cannot incorporate many of the factors that make for human wisdom and inform good decision making—empathy, compassion, humor, irony, a sense of justice, and aesthetics. Though it is capable of generating breathtaking images of our galaxy and the early days of our universe, the computer cannot experience beauty or awe, nor can it base any future decisions on these feelings. It is the combination of computational skills and our particular involuntary mental sensory experiences that makes us different from computers.

A Rough Draft of a Thought

To conclude this chapter, I've created a skeletal diagram of the interacting components of a thought. This simplified schematic isn't offered as the final word, but rather as a way of organizing our thoughts about the mind. My hope is that the

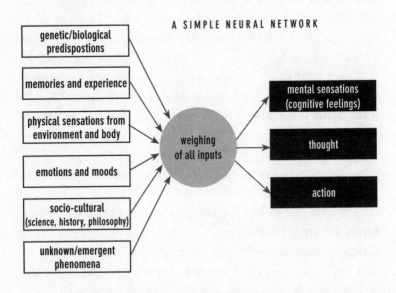

Table 1 • The Hidden Layer and Subliminal Contributions to a Thought

diagram provides a bird's-eye overview of which components of a thought seem appropriate for scientific inquiry, which might be best approached through other disciplines, and which are indeterminate.

A BRIEF SUMMARY OF MENTAL SENSATIONS

Physical sense of the self: The felt dimensions of the self, the first-person point-of-view, where the mind is located, and the personal space around the self.

Mental attributes of the self and the mind: The feelings of effort, choice, "what I am doing now," the feeling of thinking/reasoning, the sense of agency, the sense of causation, the sense of beauty, and the feeling of knowing. Each has its own constellation of closely related feelings. For example, the feeling of knowing includes the closely allied sensations of certainty, conviction, and rightness, and those that enhance the feeling, such as déjà vu, familiarity, and "real," or that detract from the feeling of knowing—feelings of strangeness, bizarreness, and "unreal" and unfamiliarity.

Some seem well localized within the brain and are easily elicited with direct brain stimulation to a region (i.e., déjà vu). Others are more complex and harder to define, such as the aesthetic sense of beauty, elegance, and symmetry. These are more likely to be widely distributed within the brain, or even composite feelings arising out of the mixing and matching of more basic sensations (i.e., the sense of reasoning arises out of feelings ranging from agency and sense of self to the sense of causation).

7 • On the Level

There were not many steps. I had counted them a thousand times, both going up and coming down, but the figure has gone from my mind. I had never known whether you should say one with your foot on the sidewalk, two with the following foot on the first step, and so on, or whether the sidewalk shouldn't count. At the top of the steps I fell foul of the same dilemma. In the other direction, I mean from top to bottom, it was the same, the word is not too strong. I did not know where to begin nor where to end, that's the truth of the matter. I arrived therefore at three totally different figures, without ever knowing which of them was right.

—Samuel Beckett, *The Expelled*

Imagine the following situation. You are the judge in a criminal trial. The defendant is an eighteen-year-old male who, with five friends, held up a grocery store. After taking the money from the elderly store owner, the six of them brutally assaulted the woman. The defendant has pled guilty, and his lawyer is trying to minimize the potential sentence by arguing mitigating circumstances. Which of the following explanations would you consider the most relevant?

1. The event was not premeditated, as it was simply a random event occurring at the quantum level. "My

defendant's electrons got out of whack and triggered an attack of violence."

2. "We have done a genetic analysis, and my defendant has genes known to be associated with an increased incidence of violent behavior."

3. "An fMRI study shows decreased activity in the defendant's prefrontal motor cortex—consistent with impaired self-control as well as a higher likelihood of sociopathic behavior."

4. "The defendant has decreased spinal fluid and blood levels of oxytocin and serotonin when under stress. These are likely to lower his general level of empathy."

5. "The defendant was raised in foster homes and has a history of repeated physical and mental abuse."

6. "The defendant expresses extreme remorse. He feels that he was swept up in the moment, was guided by the others, and 'wasn't himself.' At no time did he have any personal thought of hurting the woman."

The mind isn't a specific entity; it is a placeholder for descriptions of different levels of phenomena arising from quite different types of mechanisms without obvious causal links. How we judge the defendant's behavior will depend upon which level of function we address. Descriptions of the movement of atoms tell us nothing about cause and effect at the biological-systems levels, which in turn aren't useful for ferreting out parenting and developmental factors or the role of group dynamics. At present, there is no compelling science that bridges the various levels of explanation of human behavior. Without a first-rate understanding of the causal relationships between these functionally different levels, we are reduced to speculation and personal narrative.

In the end, you are forced to rely upon your own sense of what was going on in the defendant's mind at the time of the robbery and assault. To do so you will interject your own assessment of the defendant's sense of personal self, agency, and ability to control his thoughts and behavior, and the degree to which he can resist negative peer pressure. In contemplating the latter, your brain might offer up mental images of a riot at a soccer game in England, of a Nazi rally in Nuremberg, of a violent scene from *A Clockwork Orange*, or of your favorite hero defying military orders and refusing to shoot an unarmed civilian.

However we weigh such evidence, we tend to end up assigning psychological explanations to group behavior as though outside influences affect an individual's actions but, strictly speaking, aren't part of individual minds. Rather than sensing the possibility of a group mind, we are driven by our biology—a sense of a bounded self, personal autonomy, and agency—to feel that each of us has a unique mind separate from the thoughts of others. Whether deciding on the degree of responsibility of our hypothetical defendant or grappling with the moral dilemma of what bystanders should or could have done to prevent the Holocaust, we inevitably come up against the problem of how much personal autonomy we retain in a group setting.

If the mind is a concept rather than a physical entity, might we be better off thinking of the mind in a larger dimension? It is fashionable to talk of an extended mind when discussing our increasing reliance upon machines—from cell phones to supercomputers—to augment our mental capabilities. Whether or not an external hard drive should be considered a component of your brain seems like a question of semantics rather than a real question. Storing a memory outside your body as opposed to in your brain makes the external hard drive a clear component of your memory system. Nevertheless, such machinery is more like an appendage than an intrinsic aspect of mind, just as a cane is an assistive device for walking, but isn't considered part of your brain's motor system even if it becomes part of the representational map of your arm (as in the monkey-rake experiment).

My question is more fundamental. If our individual minds have emergent properties not found in the underlying brain cells, is it possible that further properties (higher levels) of mind can arise from the collective action of individual minds? More specifically, what if there is an inherent biological component to group dynamics that makes the notion of an individual mind inaccurate or incomplete?

An essential feature of evolutionary biology is that a successful adaptation is likely to extend across as many species as is biologically practical. We have no doubt that the heart and lungs evolved as an efficient way to provide an organism with oxygen, so we aren't surprised that a similar circulatory system is shared by much of the animal kingdom. In the same vein, we are increasingly aware of group behavior in other species, even those with little or no neural machinery for apparent intelligence—from ants directing traffic to termites building complex edifices. If such behavior results from biological systems present within a group but not within the individual, and this adaptive characteristic—group behavior—is widely distributed throughout the animal kingdom, wouldn't it make sense to at least consider this possibility for humans?

We usually think of ourselves as having individual minds upon which environment plays its tune. From the monkey-rake experiment to the rubber hand illusion, we can readily see where the environment actively changes the physical makeup of the brain. But examples from nature suggest the possibility of a group mind even in the absence of an individual mind. The discussion that follows is not an attempt to support a New Age or transcendental philosophy. What intrigues me, based upon evidence from the group behavior of other species, is the strong possibility that the notion of the unique individual mind will one day go the way of the flat Earth and the geocentric universe.

Exhibit A: Slime Mold

At the crossroads between plant and animal, slime molds are single-celled microscopic creatures that have the capacity to merge with one another to form a single larger organism.[1] Slime mold characteristically lives a solitary life when food supplies are adequate. When food is scarce, individuals merge with their brethren to form a giant amoeboid-like blob that is a marvelously efficient food finder. How this happens has been one of the great insights in modern biology. In his book on emergence, Steven Johnson has said, "For scientists trying to understand systems that use relatively simple components to build higher-level intelligence, the slime mold may someday be seen as the equivalent of the finches and tortoises that Darwin observed on the Galápagos Islands."[2]

For over half a century scientists have known that the individual slime mold cells can communicate through the release of a chemical, cyclic adenosine monophosphate (cyclic AMP). Initially it was presumed that some cells were in charge of this process, acting like pacemakers in the same way that certain cells in the heart control the rate of contraction of all the heart cells. But no such pacemaker cells were discovered; all slime mold cells are interchangeable. Eventually the pacemaker theory was dropped in favor of a bottom-up, no-one-in-charge emergent behavior. As the result of fifty-plus years of research spanning multiple disciplines—from mathematics and computer science to embryology and physics—the slime mold has become a model of "intelligent" group behavior emerging from an organism that has no nervous system with which to be "intelligent."

One could argue that food-seeking behavior is hardly intelligence. So what exactly are the intellectual capabilities of this blob of mucoidlike substance slithering along the forest floor? For some years it has been known that slime mold is capable of solving complex mazes in order to find food.[3] It does so by spreading out its network of tubelike legs (pseudopodia), simultaneously exploring alternative routes until it finds the optimal path to reach the food. It then consolidates into a single blob, taking the shortest route to the food. To further study its problem-solving abilities, two British researchers put the slime mold to a simple test. They created topographic templates of the UK using a sheet of agar, then used oat flakes to indicate the nine most populous cities, excluding London. (It appears that slime mold loves oat bran.) In the "London" location, the researchers introduced a colony of slime mold and recorded the colony's feeding activity. Within a day the slime mold had extended legs connecting the various oat bran cities; the resulting "maps" mimicked the existing British intercity road network.[4] In effect, this "mindless creature" was able to accurately determine the shortest and most efficient route to the scattered bits of oat bran—arriving at the same optimal routes as highly trained highway engineers. According to the lead researcher, "This shows how a single-celled creature without any nervous system—and thus intelligence in the classical sense—can provide an efficient solution to a routing problem."[5]

Toshiyuki Nakagaki, an expert in slime mold maze solving, reproduced this study with a topographic map of Tokyo and its environs, using oat flakes to represent thirty-six cities in the Tokyo area and then putting the slime mold at the spot corresponding to Tokyo. The slime mold accurately re-created the local Japanese railway system. Nakagaki calculated that the degree of difficulty of such problem solving is equivalent to the degree of mathematical complexity necessary for a human being to maintain balance while riding a bicycle. When asked whether or not he considered the slime mold to be intelligent, Nakagaki deferred by answering that it all depends on what you mean by intelligence.[6]

Exhibit B: The Way of the Locusts

A second example of emergent behavior that raises the question of what a "mind" is in a group setting is the swarming behavior of locusts (a type of grasshopper). Living primarily in dry areas, locusts characteristically lead isolated and relatively antisocial lives, shying away from contact with other locusts and surviving off a limited plant diet. However, periodically, when the rains come and vegetation becomes more abundant, they breed and the population soars. As long as the food remains plentiful, locusts continue with their solitary ways. When the rains stop and the land dries out, the hungry locusts crowd together into those diminished areas of remaining vegetation. This close contact triggers a dramatic and rapid change in the locusts' behavior. They begin marching together and actively seek the company of other locusts. Soon they are eating anything in sight, including each other. Within hours the locusts are transformed from solitary, finicky plant-eaters to marching, swarming, marauding cannibalistic devourers of their brethren. In a moment of possibly self-referential understatement, the academic community describes this transformed locust as "gregarious."

By examining a group of desert locusts placed in a several-square-foot enclosure, Australian researchers have found a tipping point at which this transition occurs. At low densities the insects are unorganized and go their separate ways. When the numbers reach ten to twenty-five, the locusts are closer together but still remain disorganized. However, at a critical density of about thirty locusts within the enclosure, an extraordinary set of physiological changes quickly ensue. The locusts change color, from brown to yellow and black. Their leg muscles enlarge and begin marching movements synchronized with nearby locusts. Their brain increases in size by 30 percent and undergoes a fundamental reorganization: the visual-processing areas necessary for solitary food finding are reduced, and areas that provide higher-level visual processing needed to cope with group foraging are increased.[7] (Given these dramatic changes, it's no wonder that until the 1920s, solitary and gregarious locusts were considered separate species.)

A decade ago it was discovered that this shift from solitary to gregarious behavior could be initiated by tickling a tuft of hairs located on locusts' hind legs—the same region that comes into contact with other locusts when they are in close proximity. More recently researchers have found that stimulating these hairs triggers a sudden outpouring of brain serotonin—three times the levels found in solitary locusts. Serotonin is a potent neurotransmitter; among its many brain functions is the regulation of mood, anger, aggression, and appetite. Blocking the effect of serotonin prevents the swarming; infusing serotonin into a solitary locust transforms it into a swarming monster. "Here we have a solitary and lonely creature, the desert locust. But just give them a little serotonin, and they go and join a gang," said the lead investigator, Malcolm Burrows of Cambridge University.[8]

Imagine a cartoon strip of a locust couple talking over dinner after the husband has undergone this transition. "What's come over you?" she asks. "You've always been a thoughtful, reserved, eco-conscious vegetarian, and now, look at you—even your coloring is different. I don't know you anymore." The male locust shrugs, and flashes a look of contrition. Before he can reply, his attention is diverted by the sight of a cloud of locusts hovering outside the dining room window. He gets up and begins marching toward the door. "I'll be home later. Don't wait up for me." In the next panel, Mrs. Locust is standing at the window, watching her husband join the swarm. In the last panel, Mrs. Locust has opened the window and is shouting at her husband. "Wait. I'll be right there. I've changed my mind."

Groupthink

As we don't tend to think of locusts as autonomous with a well-developed mind, we don't fret over whether the locust mind is confined to the individual. It's easy to accept that overcrowding leads to a partial rewiring and reorganization of the locust's brain. But what, if anything, does this tell us about human group behavior? Might there be similar biological effects underlying fraternity hazing practices that end with unintended deaths, the Rwandan genocide, the abuse of prisoners at Abu Ghraib, the My Lai massacre in Vietnam? Or do psychological explanations suffice? We hate being crowded, have a strong sense of when our territorial rights have been violated; too much background noise frays the nerves. . . .

Ian Couzin, a mathematical biologist at Princeton University and the University of Oxford, has found some laboratory-based behavioral evidence for human swarming, yet describes us as mediocre swarmers.[9] At the same time, based upon extensive mathematical modeling of swarming in a variety of species, he sees a comparable behavior in individual brain cells. He uses the example of basic perception—how your brain can make sense of a flood of signals coming from the eyes.

"How does your brain take this information and come to a collective decision about what you're seeing?" For him, the answer may lie in an inner swarm at our cellular level—cells communicating in a manner similar to locust-locust interactions.[10]

That we are only mediocre swarmers in lab studies isn't surprising. Rather than representing a fundamental difference in brain function between ourselves and locusts, our apparent lack of easily demonstrable swarming behavior may simply reflect the extent to which we can exert some degree of conscious control over some biological phenomena. A snapshot of a locust's hidden layer is unlikely to reveal strong cultural and moral biases against cannibalism and general mayhem. Group biological changes in lower life-forms are more likely to have a direct correlation with final behavior. As nervous system complexity increases and self-awareness arises, embedded moral, social, and cultural values as well as the conscious determination to go against one's basic instincts make the behavioral effects of group biological changes far less predictable (assuming that we have any ability to mitigate our innate tendencies).

Given the enormous complexity of human behavior and the lack of precise correlation between any suspected biological changes at a group level and individual behavior, the challenge of demonstrating a biologically based human "group mind" is truly daunting. For example, suppose you wanted to study whether watching violence on TV increases the incidence of teenagers getting into fights. You assign a control group to no TV watching and have the study group watch three hours a night of mixed martial arts, reruns of *Bronson*, the *Friday the 13th* series, and *The Texas Chainsaw Massacre*. After three months you find the same incidence of fights in the two groups. Can you conclude that TV violence didn't make a difference? You might reasonably counter that not being able to watch the programs so angered those in the control group that both groups had more altercations than expected. Until such time as we have complete understanding of all variables affecting behavior and how they are interrelated, we cannot know whether or not the supposed control group is truly a neutral control, or is being affected by a bias not recognized or considered. Given the inherent problems of determining an ideal control group for the study of complex behavior, it's no wonder that behavioral studies remain an inexact science.

I suspect that we've all had at least one locustlike experience. Imagine yourself in a packed stadium or theater. The event is over and you start toward an exit. You can't see more than a few feet in front of you and settle into following those immediately in front of you. Without any thought, you find yourself making tiny marching movements as you "feel" your way toward the exit. You don't give this altered stride a second thought; you brush off the small steps as being the only way to move forward without stepping on someone. Given your sense of agency, you feel that your altered gait is deliberate. But was this your choice or the result of a group mind, and how would you know?

Consider something as simple as keeping time to music. When tapping a rhythm by yourself, you can feel both being in control of your tapping and being carried along in a groove dictated by the band. If you're in a crowd clapping in unison with others in the audience, you have no doubt that you are willfully clapping—no one else is moving your hands for you—and simultaneously sense that you are part of a group rhythm. It was once believed that this group behavior was best understood as having an initiator or leader with others following along and "syncing up" (the pacemaker theory). To test this hypothesis, neuroscientist Chris Frith and colleagues asked volunteers to pair off and try to tap a simple rhythm together. Each subject was equipped with headsets so that he could hear the other but not himself. In this setting, no leader emerged; both players continually adjusted their beat to each other. This continuously adapting interchange between the two subjects—rather than one following the leader—can readily be seen in watching two first-rate jazz musicians improvising. No leader, no follower, no individual agency at work—just two constantly interacting members of a single unit. Frith believes that these two people are best seen as a single complex system rather than two systems interacting. Both brains are acting as a single complex unit.[11]

When interviewed, participants in such synchronization studies describe widely varying degrees of agency, ranging from a total loss of control to a heightened sense of control over the timing of their motor movements. In addition, some describe feeling controlled by the group, while others describe a shared sense of control.[12] Earlier, we saw that an altered sense of agency can occur with mental illnesses such as schizophrenia, and can be induced by suggestion, as with hypnosis. Synchronization is an example of how a behavior as elementary as group clapping can transform one's personal sense of control. It is hard not to speculate on how ubiquitous this phenomenon is, and how easily it can be used for crowd manipulation. Watch a cheerleader warming up a cheering section, a Marine drill sergeant shouting marching orders, or a great orator, and you get a sense of how easy it is to be swept up into the mesmerizing rhythm and cadence of someone with an agenda.

A sense of agency is not the only mental sensation that can be affected by societal influences. Multiple fMRI studies have shown that a localized area of the brain—the ventral medial prefrontal cortex—will show increased activity when we have thoughts and mental images about ourselves. For Westerners, this region is primarily activated when the subject is presented with self-referential words and images or actively thinks about him- or herself. For Chinese and Japanese people, activation occurs both with thoughts of one's self and when one thinks about or is presented with information or images of close family members, especially mothers. It is as though the very sense of self—at least as reflected on fMRI scans—varies according to culture. How we interpret this finding is itself a reflection of our own underlying culturally shaped brain circuitry. What in Westerners might elicit Freudian interpretations and accusations of being a "mama's boy" will in Asians be seen as evidence of family reverence and respect for tradition.

To see how this finding might play out in bicultural subjects, Westernized Chinese living in Hong Kong were tested after being first immersed in either Western or Eastern culture. On one occasion, they were presented with a series of Western images covering various cultural domains—food and drink, music and arts, popular movie stars, religion and legend, folklore, and famous monuments. After looking at each of these images for ten seconds, the subjects underwent fMRI scanning during which they were presented with references to or images of themselves and close others. The following day they underwent similar priming, this time with Chinese references substituted for the Western references.

The result: subjects' sense of self was dramatically different depending on whether it was preceded by exposure to Western or Eastern cultural icons. When Easterners were primed with Western images, the sense of self was restricted to personal references. When presented with Eastern iconography, this sense of self expanded to others. In some this not only extended to those to whom the subjects were intimately related, but even to nonrelated people in positions of authority such as their employers. In short, according to the researchers, the neural substrate for a sense of self was affected by cultural priming.[13]

A third mental sensation affected by culture is our sense of

Table 2 • Müller-Lyer Illusion

certainty. Take a look at the Müller-Lyer diagram and ask yourself if the top line and the bottom line are the same length. Even when you measure the two lines as being the same, it is hard to shake off the feeling that the lower line is longer. In recent years I have used the Müller-Lyer optical illusion to demonstrate that an intellectual understanding that the two lines are equal in length is separate from the felt sense that the two lines are of unequal length. For me, this is an argument in favor of the feeling of knowing being separate from an intellectual understanding. It never occurred to me that this cognitive dissonance generated by a basic visual perception might have cultural roots. However, in 2010, a University of British Columbia research team headed by psychologist Joseph Heinrich showed that different cultures perceive this illusion differently.

Heinrich's team showed the illusion to members of sixteen different social groups including fourteen from small-scale societies such as native African tribes. To see how strong the illusion was in each of these groups, they determined how much longer the "shorter" line needed to be for the observer to conclude that the two lines were equal. (You can test yourself at this website—http://www.michaelbach.de/ot/sze_mue lue/index.html.) By measuring the amount of lengthening necessary for the illusion to disappear, they were able to chart differences between various societies. At the far end of the spectrum—those requiring the greatest degree of lengthening in order to perceive the two lines as equal (20 percent lengthening)—were American college undergraduates, followed by the South African European sample from Johannesburg. At the other end of the spectrum were members of a Kalahari Desert tribe, the San foragers. For the San tribe members, the lines looked equal; no line adjustment was necessary, as they experienced no sense of illusion. The authors' conclusion: "This work suggests that even a process as apparently basic as visual perception can show substantial variation across populations. If visual perception can vary, what kind of psychological processes can we be sure will not vary?"[14]

Challenging the entire field of psychology, Heinrich and colleagues have come to some profoundly disquieting conclusions. Lifelong members of societies that are Western, educated, industrialized, rich, democratic (the authors coined the acronym WEIRD) reacted differently from others in experiment after experiment involving measures of fairness, antisocial punishment, and cooperation, as well as when responding to visual illusions and questions of individualism and conformity. "The fact that WEIRD people are the outliers in so many key domains of the behavioral sciences may render them one of the worst subpopulations one could study for generalizing about Homo sapiens." The researchers found that 96 percent of behavioral science experiment subjects are from Western industrialized countries, even though those countries have just 12 percent of the world's population, and that 68 percent of all subjects are Americans.

Jonathan Haidt, University of Virginia psychologist and prepublication reviewer of the article, has said that Heinrich's study "confirms something that many researchers knew all along but didn't want to admit or acknowledge because its implications are so troublesome."[15] Heinrich feels that either many behavioral psychology studies have to be redone on a far wider range of cultural groups—a daunting proposition—or they must be understood to offer insight only into the minds of rich, educated Westerners.

Results of a scientific study that offer universal claims about human nature should be independent of location, cultural factors, and any outside influences. Indeed, one of the prerequisites of such a study would be to test the physical principles under a variety of situations and circumstances. And yet, much of what we know or believe we know about human behavior has been extrapolated from the study of a small subsection of the world's population known to have different perceptions in such disparate domains as fairness, moral choice, even what we think about sharing.[16] If we look beyond the usual accusations and justifications—from the ease of inexpensively studying undergraduates to career-augmenting shortcuts—we are back at the recurrent problem of a unique self-contained mind dictating how it should study itself.

The idea that minds operate according to universal principles is a reflection of the way we study biological systems in general. To understand anatomy, we dissect one body as thoroughly as possible and draw from it a general grasp of human anatomy. Though we expect variations, we see these as exceptions to a general rule. It is to be expected that we see the mind in the same light. One way to circumvent this potentially misleading tendency to draw universal conclusions whenever possible is to subdivide the very idea of a mind into the experiential (how we experience a mind) and the larger conceptual category of the mind—how we think about, describe, and explain what a mind is. What we feel at the personal (experiential) level should not be confused with what a mind might be at a higher level—either as a group or as an extended mind. Earlier I quoted John Searle as rejecting the idea of an extended mind because it didn't appeal to common sense. The confusion in his argument arises from allowing his personal experience of common sense to shape his broader view of what a mind might be. After all, common sense is merely the strong sense of what is familiar and right, not a truth or guarantee of fact. I suspect that this is the default position for the vast majority of us. One of the problems of conceptualizing an extended mind or a group mind is that we lack an adequate mental image to counteract our palpable sense of an individual mind.

One possible solution is to approach thinking about the mind as an extension of how we think about cellular communication in general. At the physicochemical level brain cells communicate with each other by secreting various neurotransmitters that stimulate receptors on other cells. This concept is basic to our understanding of how the brain works. In a practical sense, the flow of neurotransmitters is the flow of information; at any instant our thoughts and actions are the sum total of myriad inputs onto our brain receptors. This general schema applies to all incoming information. If we are listening to the news on the radio, the information is packaged and transmitted as radio waves; our ears and auditory systems act as receptors.

Conceptually, as a mind emerges from the purely physicochemical brain to exist at a higher level, it is involved in processing information rather than neurochemicals. Think of this higher-level mental activity (the processing of information) as functioning via metaphoric receptors. The brain receives incoming information by incoming data "stimulating" brain receptors to receive and process this information. If we are planning a vacation on Mars, we can Google flight times and scout out the best resorts. The information exists on

Google servers as bits of data, is projected onto a satellite dish, then is beamed onto your home wireless connection and into your computer, where it is converted to an optical image that strikes your retina. We can accurately trace and analyze the movement of this information along this physicochemical dimension. Though we have an excellent idea of the mechanics of how information is transmitted and a growing understanding of how it is physically stored in the brain, we are woefully ignorant of what information actually is at a physical level. The entire field of information theory struggles with this essential mystery. At the next level of thought up, metaphysicians debate the existence of platonic ideals, fundamental truths, and moral laws—how seeming verities exist outside the overtly physical dimension.[17]

This problem is common to many if not all physical properties. Take gravity. We have quite elegant mathematical descriptions of the behavior of gravity, yet no one knows what gravity is or the state in which it exists. Gravity is presently known to us only by its effects, not by any direct observation independent of its effects. Quantum theorists argue that gravity represents some as-yet-to-be-discovered subatomic particle; Einstein claimed that gravity is an integral quality of the fabric of space-time. Irrespective of what gravity might one day turn out to be, or whether it will ever be uncovered at a fundamental "what it is" level, we can all agree on its effects. Similarly, a wide range of higher-level phenomena such as cultural values and group dynamics clearly cause biochemical and structural brain changes. That information exists and affects how we think, even how our brains are wired, is beyond doubt.

Once we see the conceptual mind as receiving information, the sky is no longer the limit. Complexity theorists tell us that a butterfly's fluttering wings in Tokyo can cause a sandstorm in Timbuktu. Quantum physicists beat the theoretical drum over entanglement—the measurable interaction of electrons situated at opposite ends of the universe.[18] The observation that groups of brain cells seem to have their own version of quantum entanglement, or, as Einstein put it, "spooky action at a distance," has prompted some neuroscientists to believe that this might explain how our minds combine experiences from many different senses into one memory.[19] If quantum entanglement can be seriously entertained, spooky action at a distance might also apply to the ability of information, wherever present, to affect the individual mind.

As we go forward, each of us has to find his or her own way of balancing two entirely separate ways of seeing the mind. Having read the above paragraphs, you are still going to feel a strong sense of a personal presence of a mind constrained by the dimensions of your sense of self and endowed with causal powers. At the same time, a scientifically sound operational understanding of the mind requires an acceptance that higher levels of mental activity such as the receiving of information might extend far beyond the confines of our individual brains and bodies.

Perhaps the above argument sounds like a frustrating exercise in semantics. Whatever happens out in the world is still manifested by physical changes within our brains. But to limit our vision to individual minds is to limit our study of the mind to individual effects. Imagine wanting to study how close proximity can affect neural circuitry. What would be a good experimental design? Would it make sense to put a group of subjects in adjacent fMRI scanners and ask them questions all at once? Show them pictures of overcrowding? But what if we had taken that approach during the field study of locusts? Unless the locusts physically rubbed up against each other, we would not have seen the biological changes triggered by the crowding. How accurate will our understanding of empathy be if we study solitary individuals for brief periods while they are isolated in scanners rather than in close physical proximity with others? If studies on priming are correct, bicultural students coming from a class on Western civilization or Asian studies may have significantly different fMRI responses on cognitive testing. Consider how the response of two subjects might vary simply because one has spent the night watching Bruce Lee movies and the other James Bond reruns. The obstacles are profound, the lessons for caution in interpretation undeniable.

It is hard to imagine how to obtain a truly neutral and well-controlled fMRI study in which all such subtle confounding variables have been eliminated. To do this, we would have to know all of the various effects of all preexisting activity on all aspects of brain function—a task of inconceivable proportions. If we wish to understand such phenomena as group behavior, cultural biases, or even mass hysteria, it seems preferable to see the mind in its largest possible context rather than persisting with the arcane notion of the individual mind under our personal control. The receptors of our conceptual mind reach out to the far corners of the universe even as our experiential mind tells its personal tales and sings its unique songs just behind our eyes.

Wild Speculation

Ever since reading about slime mold's maze-solving capabilities, I have been haunted by the possibility that human intelligence also might have a biologically mediated group component. When I watch a school of fish, a flock of starlings putting on an aerial show, a pod of whales collectively beating the waters to round up their dinner, I cannot help wondering if similar mechanisms contribute to partisan politics, corporate groupspeak, group conformity and reluctance to consider new ideas, even heroic behavior as seen in World War II resistance groups. Unless we are truly different from the rest of the animal kingdom, it's quite likely that we possess similar basic biological mechanisms. At the same time, it's quite unlikely that these mechanisms can be neatly bundled into precise explanations for particular human behaviors. If we were to find increased serotonin levels in the brains of rock concert spectators, we would still be left speculating as to the underlying reasons. The complexity of the human nervous system and of the human experience doesn't allow for a foolproof way of sorting out the possible contributing causes. As we can see from the TV-violence argument, humans, unlike unicellular slime mold or small-brained locusts, aren't amenable to precise delineation of their behavior by the manipulation of a single variable. Even if we were to find hairs on the back of our legs that jacked up serotonin levels when stroked, we wouldn't know if the stroking was the direct cause of the increased serotonin levels. Perhaps stroking a subject's leg relieved an itch or triggered a fond memory, which in turn increased the serotonin levels.

At the risk of sounding utterly nihilistic, there is a second major drawback to studying the underlying biology of group behavior: our fundamental lack of knowledge of brain function at a cellular level. For over two hundred years it has been known that there are two major types of brain cells: the neurons that presumably carry out our thinking, and other "stuff." The latter—glial cells (from the Greek word for "glue," as it was long thought that these cells held the brain together)—come in several varieties. One class, the oligodendrocytes, is responsible for generating the insulation (myelin sheaths) around nerve fibers (axons). Another—the astrocytes—is integral to neuronal function, providing nourishment, cell regulation, even blood vessel control in the brain's microcirculation. Until very recently it was generally accepted that neurons make thoughts, while glial cells are the supporting cast that make the star neurons flourish. But this view of glial cells may be undergoing dramatic revision.

A bit of history: In the late nineteenth century, a Spanish neuroscientist and Nobel laureate, Santiago Ramón y Cajal, developed elegant staining techniques that allowed detailed observations of our neurons and their interconnections. Ramón y Cajal is considered by many to be the father of modern neuroscience; his work is credited with popularizing the prevailing belief that neurons are responsible for our thoughts—the so-called Neuron Doctrine. This view was further solidified by technological developments in the 1930s: the isolation of squid axons large enough to be studied electrically with intracellular recording electrodes. By the mid-1940s, British scientists Alan Hodgkin and Andrew Huxley had determined the nature of nerve impulse transmission—the electrical action potential that travels the length of the nerve and results in the release of neurotransmitters into the synaptic cleft. Their work, also garnering a Nobel Prize, is fundamental to our understanding of the nervous system.

Meanwhile, despite accounting for half the volume of the adult mammalian brain, and being at least as plentiful as neurons, glial cells received scant study. Harder to investigate, they remained on the sidelines of neuroscience. It wasn't until the 1960s that the astrocyte was also shown to have action potentials. Then it was discovered that both neurons and astrocytes responded to and released neurotransmitters. Most recently it was found that astrocytes can also create calcium waves that extend to an area a hundred times larger than the responsible astrocyte. Though astrocytes don't have their own synapses, a significant percentage of their end plates (up to thirty thousand per astrocyte) are closely approximated to neuronal synapses. All the features for them to affect neural transmission are in place. But are they responsible for, or do they contribute to, cognition?

The verdict is still out. Some experts believe they play no role in cognition. Others are convinced that they play some role, but aren't sure of the extent of this contribution. Some remain undecided. In 2008, University of North Carolina researcher Ken McCarthy wrote that "astrocytes may be active participants in brain information processing."[20] A year later, his next experiment failed to reveal convincing evidence of glial effects on neurons, causing McCarthy to question his earlier position. Those most unreservedly enthusiastic opine that glial cells are a major factor in the generation of our thoughts. In a recent interview in *Scientific American,* University of Wisconsin neuroscientist Andrew Koob said that "Astrocytes control neurons, not vice versa." "It is obvious that astrocytes are involved in brain processing in the cortex, but the main questions are, do our thoughts and imagination stem from astrocytes working together with neurons, or are our thoughts and imagination solely the domain of astrocytes?"[21]

Does Size Really Matter?

To bolster his argument, Koob relies on prior neuroanatomical studies of cell density within the brain. In the 1960s, it was estimated that glial cells accounted for nearly 90 percent of brain cells. Koob cites this observation as the origin of the popular myth that we use only 10 percent of our brains. Invoking our tendency to assign importance to size and quantity, as in "big is better" and "the more the merrier," Koob's implicit argument is that the greater the number of glial cells, the more likely they are to play a primary rather than a mere supporting role. But cell count studies—and thus, one presumes, the potential importance of glial cells—vary widely; at least one earlier study has suggested a fifty fold preponderance of glial cell to neurons.[22] And new techniques bring new results. A study published six months before Koob's *Scientific American* interview in 2009 suggested a ratio of approximately one neuron to one glial cell.[23]

This variability of cell counts also applies to neurons. Estimates of the total neurons in the brain vary dramatically, ranging from ten billion to one trillion neurons.[24] It might seem surprising that in this day and age of techniques sophisticated enough to be able to unravel the human genome, we cannot accurately count the number of brain cells, but there it is. Different techniques bring different results. It is hard to know whether any of the present values will stand the test of time.

According to Swiss researcher Andrea Volterra, the stakes are enormous. "If glia are involved in signaling, processing in the brain turns out to be an order of magnitude more complex than previously expected. Neuroscientists who have long focused on the neuron would have to revise everything."[25] But there's a hitch. Not only is the verdict out, but it seems that the jury is deadlocked over how to proceed. No one has been able to come up with an experimental approach that can be agreed upon as a potential source of a definitive solution. In a 2010 review in *Nature*, University of London neurobiologist David Attwell said, "There's no clear simple experiment, otherwise I'd do it. . . . So would most of the others."[26]

Try to think of an experiment that would yield unequivocal results. Suppose we wanted to see the nature of our thoughts (if any) if we didn't have glial cells. As glial cells are integral to neuronal function, it is impossible to design a study in humans in which glial cells are inactivated. Without glial cells, neurons wouldn't work properly. Even selective inactivation of a component of glial cell function would raise the same is-sues. If neurons were, for example, the messenger for thoughts generated by glial cells, we would see nothing and be able to conclude less. In addition, at present we have no way to sort out what emergent properties might be present in a collection of glial cells. In effect, we would have to build a functional brain composed only of glial cells, hook it up to a body, and then try to study its effects.

For me, the most important takeaway in this glia-neuron story is the role of scientific methodology in generating hy-potheses. In large part because neurons were easier to study experimentally, we have had the prevailing belief that neurons are the primary source of cognition. Meanwhile, until recently, science has largely ignored a more difficult to study brain cell of equal (or greater) numbers, brain volume, and anatomic distri-bution. If glial cells had been easier to study than neurons, we might have an entirely different understanding of how the brain creates a mind. Maiken Nedergaard, a glial biologist at the Uni-versity of Rochester, attributes much of the difficulty in under-standing glia to cultural bias. "The entire field has been trained in neuron-centric labs, and everybody has so far believed that astrocytes work like neurons. But astrocytes function totally differently. They use a different language. They use a different way of getting input and output. They may also work in a to-tally different timescale from neurons."[27]

How we think about the brain creating a mind arises out of our tools. I have brought up the subject of glial transmission because it raises a serious question about the evolution of our present conception of how the brain works and how thought is created. I do not know whether or not glial transmission will ultimately be shown to be important. Worse, given the honest appraisal by many of the leaders in the field, it seems unlikely that the problem can be adequately addressed with our present tools. It may be that the complexity of neuronal-glial interaction will defy complete elucidation for the foreseeable future. Meanwhile, we are left with speculation and conflicting evidence. How we integrate this imprecise and ambiguous information will depend on our mind-set, which itself will be affected by group and cultural influences on our perceptions.[28] To brush aside these inherent methodological difficulties is the same category of mistake as believing that we are close to understanding dark matter, dark energy, the nature of gravity, and a thousand other big-league mysteries that have defied the best minds and the latest and greatest technologies. In cosmological circles it is said that the visible universe is but a fraction of what is out there. The rest—dark matter and dark energy—is the subject of a mixture of experimentation, speculation, hypothesizing, and good old-fashioned science fiction. It is hard to avoid seeing the brain in the same light. Perhaps glial cells should be seen as dark brain matter as opposed to white matter.

A What-if

To show how hard it is to step away from our deeply held beliefs, consider the following hypothetical. If group "intelligence" is seen throughout the animal kingdom, is it possible that we have in place a similar cellular system for complex "intelligence"? Take a moment to ask yourself how you feel about the idea of your mind being partially driven by biological mechanisms that might extend beyond your individual brain. Is this exciting, depressing, nonsensical, alarming, titillating? Would this affect your sense of self-worth, moral values, relationship to others, religious beliefs? How you think about this possibility is intimately related to how you experience your own mind—your attributions of agency, sense of self, personal uniqueness. It is also related to how you see yourself in relationship to the rest of the animal kingdom, determine the validity of an idea in the absence of compelling evidence, and feel about your ability to pick and choose among differing expert opinions; and whether you go along with prevailing cultural and scientific beliefs, and believe that science will eventually unravel all the mysteries of the universe. The list of influences is virtually infinite.

Though it's possible to infer it from the behavior of other species, hard evidence for a group mind is both lacking and likely to remain out of reach for the foreseeable future. In thinking about the mind's limits and dimensions, the metaphors of information as neurotransmitters and the mind as a receptor are appealing to me. So is the idea of group behavior partially being under shared biological influences. I cannot prove that these ideas are correct, but even the possibility allows us a broader view of what neuroscience can say about the mind.

In summary, at the neuroanatomical level, the dimensions of a mind are strikingly indistinct. Despite the most sophisticated of techniques, we aren't sure how many cells there are, let alone how they interact. A biologically mediated group impact on a thought dramatically increases the possible degree of complexity yet isn't an easy target for investigation. There are huge gaps in our understanding of other pertinent aspects of basic science—from emergence/complexity to quantum entanglement. Involuntary mental sensations help drive and judge our self-observations and our investigations into the mind. These issues are but a few of the hurdles with which neuroscience is faced.

8 • Talking in Tongues

The original is unfaithful to the translation.

—Jorge Luis Borges, concerning the
Vathek by William Beckford

The recent surge in the public interest in neuroscience is largely driven by the hope that scientific investigation can provide us with a better understanding of human nature than previous psychology-based theories. But the basic language of present-day neuroscience can't provide this understanding. It does us no good to know, as described in a recent journal article, that a person is experiencing "increased activity in regions of the dorsal anterior cingulate cortex, the supplemental motor area, anterior insula, posterior insula/somatosensory cortex, and periaqueductal gray and that the temporo-parietal junction, the paracingulate, orbital medial frontal cortices and amygdala were additionally recruited, and increased the connectivity with the frontal parietal network."[1]

Such professional jargon is essentially incomprehensible and meaningless to all but a select few. Just as we need a translator to tell us what's written on an ancient Sanskrit papyrus, neuroscientists must translate their arcane texts into a readily understandable common language. They must tell us that these regions of the brain constitute a pain matrix and that these neural structures are involved in our personal experience of pain. As a consequence, neuroscientists must wear two separate hats—investigator and translator. They ply their trade, for which they have been trained, and then take on the second role of translator and explainer of their own data. Unfortunately, this brings us back full circle; neuroscientists must translate their findings into the language of popular psychology.

The inherent difficulties in assuming this dual role of experimenter and translator cannot be overstated. Basic neuroscience is a highly complex and difficult field; most neuroscientists have a relatively narrow field of real expertise. With tens of thousands of cognitive scientists churning out new information, keeping up to date is a monumental task. For basic scientists to also be well informed in psychology is impossible. Not having the time, and often lacking the background, training, or interest, they must, to explain their findings, rely on popular psychological theories that they are often ill-equipped to judge. Experimental psychology is a field unto itself. Years of study are necessary in order to achieve even a superficial understanding of the innumerable pitfalls of experimental design and interpretation.

Similarly, psychologists, cognitive scientists, and philosophers increasingly incorporate summary conclusions of neuroscience to support their ideas, but without having the training to recognize inherent limitations of basic science methods and interpretations. The cycle is never-ending. New psychological theories become the neuroscientists' language for translation of their own basic science data, which in turn are cited by the psychologists as evidence for their theories. Once an idea gets a foothold in the collective mind of the cognitive science community, it develops a life of its own, irrespective of its underlying validity. Unsubstantiated word-of-mouth morphs into hard fact.

To give a sense of the inherent limitations of translating hard data into the popular vernacular of folk psychology, I've chosen a few high visibility subjects to consider in the next few chapters. My goal isn't to pick apart individual findings, as this isn't particularly helpful in a broader context. Nor is it to launch personal attacks on the mostly well-meaning scientists. Rather, I'd like to offer a more practical way to assess the quality of any neuroscience claim. In so doing, I have picked out articles that are likely to have significant influence on our future understanding of aspects of behavior, ranging from empathy and intelligence to free will and determination of consciousness. My goal isn't to refute the observations but to question the degree of confidence in the conclusions. Let's begin with the discussion of mirror neurons.

Reflecting on Mirror Neurons

In the late 1980s, Italian neuroscientist Giacomo Rizzolatti and colleagues were studying the premotor region of the frontal lobe of a macaque monkey. Using intracellular electrodes, they were able to locate and record the electrical activity of individual cells that fired when the monkey reached for bits of food. As the story goes, one monkey was resting between studies, intracellular electrodes in place, just watching his experimenters. When one of the researchers reached out and picked up a peanut, the same cells began firing, as though the monkey were reaching for the food. Rizzolatti painstakingly determined that this region of the brain contained certain cells that would fire when the monkey performed a single highly specific hand action such as pulling, pushing, tugging, grasping, picking up, or putting a peanut in its mouth, and that these same cells would fire when observing another performing the same action. It was also noted that the movement had to appear intentional—as though the hand was reaching to grasp the peanut in order to eat it, rather than merely making the same gesture without the experimenter intending to eat the peanut. Given their combined capability of recording the observation of an action and initiating the action, these cells soon became known as "mirror neurons," and collectively as the "mirror neuron system."

It's important to consider how we learn a motor action. Imagine taking up a hobby for which you have no prior experience—cello playing. You have no idea how to hold the instrument, where it fits between your legs, how to make a sound with the bow. You painstakingly learn by observing and trying to imitate what you see. (This is true for any motor act, from crawling or walking to talking or texting.) This learning process—observation and imitation—is accomplished by the creation of neural circuitry: a representational map specific to cello playing. Each time you watch your teacher play, the cello circuit is enhanced. Each time you practice, it is enhanced. If you had electrodes placed within your cello circuitry, you would be able to see increased activity under both conditions. Learning to play an instrument is the act of trying to synchronize what you've observed with what you are actually doing.

Let's throw in a few more details. Your teacher's cello is old and has a wonderful fragrance. As she takes it out of its case, you are vividly reminded of that high school field trip to hear your first concert. Afterward, you were taken backstage and given a demonstration of the various string instruments. You remember holding several very old violins in your hand, even sniffing them and envisioning what it would have been like to play them when they were new. Your teacher quickly snaps you out of this brief reverie by playing an exquisite passage from a Bach prelude. To your surprise, and completely out of character, you well up with tears. You compliment her, but instead of being gracious, she sternly reminds you that when she was your age, she practiced eight hours a day. Your tears evaporate as you shift gears and wonder how anyone could devote her life to rubbing dried horsehair across a few strands of catgut when she could be out playing ball or tweeting her friends.

Now imagine this brief scenario taking place while you are wearing a portable fMRI scanner. Most of us would expect to see increased activity in areas of the brain involved in motor performance, observation of another's motor actions, processing of smells (olfactory regions), areas active in memory storage and recall, and the experience of vivid emotions. Any complex neural circuit, whether it represents cello playing or Proust tasting his famous madeleine, functions by coordinating the activity of a number of brain areas. Lumped together in this circuitry are both our observations of an action and the acquired motor skills to carry out this action.

A simplified schema for learning an action: Observation → Detailed Template (representational map) that is stored in memory. Over time, as learning proceeds, the neural substrate for the observation and the action merge into a single observation/action representational map necessary to carry out the learned action.

Rizzolatti's finding of a combined observation/action system (the same neurons fire when a monkey reaches for a peanut as when it observes the experimenter reaching for a peanut) confirms at the cellular level what we already suspected at a commonsense level. What could not have been anticipated was the degree of excitement his discovery has generated. In the subsequent two decades his work has been posited as the biological basis for how we read minds and experience empathy. But how much of this speculation is justified?

By tracing the evolution of the alleged implications of mirror neurons, we can get a sense of some of the inherent problems of translating good basic neuroscience into behavioral explanations.

Shortly after Rizzolatti's findings were published, eminent

University of California, San Diego, behavioral neurologist
V. S. Ramachandran predicted,

> The discovery of mirror neurons in the frontal lobes of mon-
> keys, and their potential relevance to human brain evolution,
> is the single most important "unreported" (or at least, unpub-
> licized) story of the decade. I predict that mirror neurons will
> do for psychology what DNA did for biology: they will pro-
> vide a unifying framework and help explain a host of mental
> abilities that have hitherto remained mysterious and inacces-
> sible to experiments. . . . With knowledge of these neurons,
> you have the basis for understanding a host of very enigmatic
> aspects of the human mind: "mind reading" empathy, imita-
> tion learning, and even the evolution of language. Anytime
> you watch someone else doing something (or even starting to
> do something), the corresponding mirror neuron might fire in
> your brain, thereby allowing you to "read" and understand
> another's intentions, and thus to develop a sophisticated "the-
> ory of other minds."[2]

Let's assume that a monkey's mirror neurons can detect
the difference between an intentional act (reaching out to pick
up a peanut for the purpose of eating it) and a nonspecific but
similar-appearing hand movement. This tells us nothing
about a monkey's ability to read another's mind, be it another
monkey's or that of a researcher in the lab. Monkeys are good
at picking up peanuts. They are also good observers of other
monkeys picking up peanuts. It shouldn't be surprising that a
macaque monkey can detect subtle differences between intentional and nonintentional hand movements. But recognizing
subtle differences in motor gestures is a far cry from reading
minds. Though the research is less than conclusive, most
studies indicate that adult macaque monkeys have little if any

ability to infer the intentions of others at a level greater than simple motor movements. Even chimpanzees are quite limited in these areas.[3] If the presence of mirror neurons isn't a good predictor of mind-reading abilities in our closest relatives, is it a good indicator in man?

To highlight the difference between recognizing motor intention and real mind reading, imagine being in a high-stakes poker game. You are about to make a bet when you notice that the player to your left is also moving his hand forward, perhaps reaching for his chips. The movement is extremely slight; you are not sure whether he intends to bet but is acting prematurely, or he is intentionally trying to deceive you to prevent you from betting. Both are reasonable options. The better you are as an observer, the more likely you are to be able to distinguish a feigned forward movement of your opponent's hand from an intentional but premature movement of reaching for his chips. To do this, you will rely on your past experience. In your years of playing poker, you will have laid down representational brain maps corresponding to various hand gestures of poker players. You will draw on this information to determine the probability of deceit versus premature action and infer what the player intended.

But knowledge of intention of a motor act will provide you with little if any knowledge about the more complex mental state of the player. For example, he may be making this gesture to distract you from another aspect of the game. Perhaps he is working in collusion with another player on the opposite side of the table and wishes to direct your attention away from the other player. He might be trying to create a fake "tell" to use against you in the future. By making this gesture and then turning over a bad hand, he could be setting you up for a future hand in which he makes the same gesture but has a great hand and beats you out of a monster pot. In short, interpreting the intention of his motor act is not the same as reading the player's mind. Intention at the motor level can be created by a number of quite different mental states. Knowing the intention of the act is not the same as knowing the purpose behind the intended action.

There is no reason to believe that the monkey knows why the experimenter is reaching to eat the peanut. The experimenter may be hungry or bored, or want to see if the peanuts are stale. As one of the mirror neuron pioneers, UCLA neuroscientist Marco Iacoboni, acknowledges the mirror neuron system operates at the basic level of recognizing simple intentions and actions. "In academia, there is a lot of politics and we are continuously trying to figure out the 'real intentions' of other people. The mirror system deals with relatively simple intentions: smiling at each other, or making eye contact with the other driver at an intersection."[4] Philosopher and cognitive scientist Alvin Goldman agrees that our ability to recognize emotions of others based upon their facial expressions is an example of low-level mind reading. He points out that low-level cognitive processes are unlike high-level ones because they are "comparatively simple, primitive, automatic, and largely below the level of consciousness."[5]

Nevertheless, this leap from low-level gesture recognition to rampant speculation that a specific set of cells can read the mind of another has become popular lore. Listen to these comments by mirror-neuron researchers.

Simone Schütz-Bosbach, neuroscientist, Max Planck Institute for Human Cognitive and Brain Sciences: "Understanding others' actions is a key function in social communication. Re-enactment through mirror neurons probably helps us to understand what another person is doing and why, and most importantly, what the person will be doing next."[6]

Ramachandran (referring to mirror neurons): "We are in-tensely social creatures. We literally read other people's minds. I don't mean anything psychic like telepathy, but you can adopt another person's point of view."[7]

Iacoboni (in a striking contradiction of his earlier comment confining mirror activity to simple intentions): "With mirror neurons we practically are in another person's mind."[8] "Neu-ral mirroring solves the 'problem of other minds' (how we can access and understand the minds of others)."[9]

Once mirror neurons were shown to be able to detect the intentions of another's actions, and, by inference (incorrect), to give one the ability to read another's mind, experiencing empathy might seem like a logical next inference. The argu-ment is that if we can put ourselves in the mind-set of another, we are better able to empathize with him, or in the vernacular of the day, "feel his pain."

To test this claim, let's take a short look at the present-day understanding of empathy. But first, let me make the distinc-tion between the ability to intellectually understand another's mental states—"You look sad"—and affective empathy, in which we actually experience another's sadness. The former is purely a cognitive and intellectual recognition of a mental state; the other is the shared emotional experience. For this discussion, I am using the term "empathy" to refer to the af-fective component—feeling what another is feeling.

One critical piece of evidence for mirror neurons playing a vital role in the human experience of empathy comes from a 2010 UCLA study on twenty-one epilepsy patients undergo-ing preoperative electrical cortical mapping. Such patients routinely undergo intracranial electrode placement (while awake) in order to identify vital areas of the brain to be avoided during the surgical removal of the area of the brain causing the patient's seizures. As some of the patients had abnormalities in the medial temporal region, neuroscientist Roy Mukamel and colleagues looked at a function well correlated with this area of brain: the recognition of facial emotional expressions. When shown a number of such facial expressions and then asked to mimic them, the patients demonstrated similar degrees of electrical activity in a small subset of neurons. Cells in the temporal lobe that responded to observed emotional responses also fired when the subjects made facial expressions. Mukamel's discovery was headlined in the media as "Empathetic Mirror Neurons Found in Humans at Last."[10]

If mirror neurons are the underlying common pathway linking imitation, mind reading, and empathy, we should expect these behaviors to be clustered together. Those who are better at reading the minds of others would be more likely to be empathetic, and vice versa. But common experience paints a different picture.

A great baseball player can watch old baseball movies and pick out subtle swing changes in former greats that are lost on a lesser batter. This same player can be mind-blind, bereft of the slightest degree of understanding of others and/or feelings for others. In this case, you could argue that he has great mirror neurons for motor actions, but that motor-mirroring skill doesn't translate either into mind-reading abilities or feelings of empathy.

Others are great at mind reading but lack any emotional empathy. Bernie Madoff comes to mind. I'm tempted to say that to have played his investors as flawlessly as he did for several decades, Madoff knew his investors' minds better than they did—presumably good evidence for a well-functioning mirror neuron system. But contrary to Ramachandran's view that mirror neurons are synonymous with "empathy neurons,"

Madoff gets a zero in the empathy department.[11] Contrast his contempt and disregard for his accusers with his savvy, sophisticated understanding of what his neighbors might expect from him, and we get a sense of the disconnect between understanding the thoughts of others and genuinely sharing their feelings. Shortly after his arrest, Madoff posted the following letter in his apartment building entranceway:

> *Dear neighbors,*
> *Please accept my profound apologies for the terrible inconvenience that I have caused over the past weeks. Ruth and I appreciate the support we have received.*
> *Best regards,*
> *Bernard Madoff*[12]

Conversely, we can feel great empathy without the slightest sense of reading another's mind. Perhaps the most compelling example is the degree of empathy we can feel for animals that we doubt have any significant consciousness or self-awareness. On a recent walk, I noticed a centipede slowly make its way around a rock. As silly as it might sound, I felt a powerful sense of connection with the centipede. I can even recall sensing the degree of effort it had to exert just to make its way across the path. The point is too obvious to belabor. Empathy toward other creatures can't have anything to do with mind reading if you don't think that the creature has a mind. (It is often easier to be empathetic when you don't know what another person is thinking.)

To challenge the notion that observation and imitation are at the root of empathy, consider how those who have never felt pain can still empathize with another's pain. French neuroscientist Nicolas Danziger studied a group of patients with congenital insensitivity to pain—a rare genetic sensory nerve disorder present at birth. People with this disorder know pain only as a concept, not as a personal experience. Interested in seeing how these patients would respond to seeing others in pain, Danziger showed them photos of a person getting her finger caught in gardening shears and a video clip of quarterback Joe Theismann's leg being broken on *Monday Night Football*.[13]

To Danziger's surprise, some of the pain-insensitive patients responded on fMRI similarly to normal controls—their pain-perception regions lit up. Others had the anticipated lack of response. Danziger found that the response neatly correlated with the degree of empathy each subject displayed on a standard empathy assessment questionnaire. Despite lacking the ability to physically appreciate the pain of another, those who scored highest on the series of questions designed to assess one's general degree of empathy had the highest degree of emotional experience of the suffering of another. The degree of empathy elicited had nothing to do with any prior experience or observation of personal pain, but, as the authors concluded, was a separate predisposing "empathy trait." By trait, the authors meant a general predisposition not accounted for by prior direct learning—an observation that is supported by a growing body of literature suggesting that the degree of one's empathy is strongly influenced by genetics.[14] If true, this would argue against the feeling of empathy primarily originating in the observation and mirroring of others.

For me, empathy is the social glue for our civilization. From rudeness and indifference to outright hostility and aggression, lack of empathy is antithetical to the well-oiled running of society. Understanding the biological components of empathy and the degree to which they can be affected through training and education is one of the great challenges at both the social and the neuroscientific level. The question of rehabilitation of the chronic criminal offender—the callous individual without remorse—will depend on whether or not we decide that empathy can be instilled, enhanced, or induced. Whether trying to figure out how to reduce political tension or to find common ground between science and religion, we inevitably end up relying on our intellectual understanding of empathy along with how strongly we feel for others (affective empathy). Premature and/or simplified conclusions about this complex problem aren't helpful. Making the unwarranted leap that empathy arises from a collection of specialized brain cells poses more problems than answers.

Wrong-Level Thinking

To design meaningful studies, scientists must begin by controlling for as many variables as possible. The smaller the scope of the project, the more accurate can be the observations. For example, in studying vision, one tries to isolate a single component such as the detection of edges or borders, or linear movement, or color. By piecing together these observations on individual components, we can draw a composite picture of how vision is generated. But this method is dependent upon having a good working knowledge of how the components function both individually and collectively. In the 1950s and 1960s, intracellular recording studies uncovered what were thought to be cells specific to particular visual functions. It was believed that some cells responded exclusively to lines, others to movement, and yet others to edges and borders. The latter were dubbed "edge detector neurons." A half century of further research has shown that this picture is far too simplistic; seeing something as simple as an edge results from the complex interaction of hundreds of cell types. There is no such thing as a specific "edge detector neuron."

The brain is a patchwork of related functions that have

evolved over aeons. Other than the most primary movements, such as the twitch of a single muscle, events such as thoughts and actions are the product of complex, widely distributed, and interrelated circuitry. There is no brain center for gratitude or remorse. The feeling of empathy has been ascribed to at least ten brain regions.[15] Though science works by looking at the smallest possible subunits, it is important not to confuse these lower-level findings with higher-level functions. Terms like "empathy neurons" mix different levels of function and action, effectively reducing the enormous complexity of brain actions to cartoonish and often misleading sound bites. No neuron causes any specific complex behavior. One cannot reduce higher-level behaviors to lower-level neuronal activities. Just as you cannot expect to read a great novel by staring at the alphabet, you can't find behavior at the cellular level.

Cells Don't Behave

A side product of assigning behavioral properties to individual types of brain cells is the mistaken assumption that the presence of particular cells is proof of the behavior. Case in point: the spindle cell. The large spindle neuron with its single axon and dendrite is found in areas of the human brain that have been implicated in emotional processing, including feelings of empathy. Originally thought to be confined to humans and great apes, they have also been found in several marine mammals, including dolphins and whales.[16] The discovery has been hailed as anatomic evidence for whales having the capability of feelings for others. The line of reasoning: Spindle cells exist in areas of the human brain that process emotions and also exist in similar regions in the whale brain. Therefore whales experience similar emotions. In effect, we are validating our observations about higher-level animal behavior, from social organization to communication, by assigning this behavior to a particular cell type.

Complex behaviors such as empathy can't be determined by the presence or absence of a particular brain cell type or a particular anatomic configuration of the brain. If this were true, we could ignore behavioral observations and skip directly to the alleged bottom line: if the brain of a particular species was thoroughly dissected and no spindle cells were discovered, we could simply write off that species as being without empathy. Nothing could be more shortsighted. Presently we have little idea of the function of spindle cells. The same argument applies to mirror neurons. Though they have been electrically isolated, there has been no histological (microscopic) confirmation that mirror neurons represent a particular cell type with a unique biochemical makeup and function.

It's true that uncovering the presence of spindle cells in other species is of great value in furthering our understanding of the brain's evolution and our relationship to other species. But the interpretation of individual cells as being responsible for complex behavior runs the long-term risk of establishing technology as the final arbiter of what another is experiencing. I can remember a time when the pejorative "anthropomorphism" was raised whenever one attributed a specific trait to another species based upon our presumption of what that animal was experiencing. There is an obvious element of truth to this criticism. We cannot know what being a bat feels like. Assigning specific behaviors to individual cells is on equally shaky ground. Rather than hope that neuroscience will bail us out of a seemingly irresolvable predicament, we are better off acknowledging that empathy arises from the brain but cannot be found within individual cells or their connections. Cells and circuits feel nothing. It is only by their collective actions and via as-yet-unknown mechanisms that we experience feelings such as empathy.

The inability to reverse-engineer behavior into its basic building blocks applies to all aspects of mental states and is the central deterrent to our understanding of consciousness. We are a long way from understanding individual neurons, and even further from understanding how they interact both within individual systems and more globally within the brain. A recent *Journal of Neurophysiology* editorial sums up our present state of ignorance: "The processes and mechanisms whereby individual neurons integrate and compute converging information from multiple sources remains as one of the more intriguing issues in neuroscience."[17]

Window of Opportunity

The path of discovery of the mirror-neuron system underscores how the very design of an experiment can have unsuspected and unwarranted long-term effects. Because Rizzolatti was investigating hand movements in monkeys, the original description of "mirror neurons" was confined to the corresponding motor regions. Had he been studying facial expressions, the mirror-neuron system would have had a different anatomic localization. It's no wonder that the extent of the mirror-neuron system is rapidly growing as other regions are being studied. At the same time as the UCLA researchers were finding mirror neurons in the medial temporal lobe of humans, Rizzolatti discovered mirror neurons in another region of the monkey temporal lobe (the insula). It is likely that as more areas are studied, the mirroring process will be regarded as a generalized neurophysiologic phenomenon widely distributed throughout the brain.[18]

I doubt that the general mirror mechanism is confined to motor actions. Consider how we learn a new idea. If you are listening to talk radio and hear a political diatribe on immigration policy, elements of the original talk will be stored as a memory. At a later date it might be delivered back into consciousness during your ruminations over the best presidential candidate. Though the original memory might be processed and stored in a different area of the brain than your presidential rumination, both will be intimately connected to each other as part of a neural network for evaluating which presidential candidate has the best stance on immigration. If we see thoughts as the mental actions of our mind, then the observation (hearing the idea on talk radio) and your new mental action (considering who's the best presidential choice) will both arise from this neural network. Of course, this process won't show up on fMRI as part of the mirror-neuron system, as it doesn't represent a physical motor action. What we will see are different areas of the brain lighting up during different types of observations and mental actions. But the same underlying general principle will apply: Observation and action will be generated by the same constellation of neurons. As mirror neuron expert Simone Schütz-Bosbach has said: "Research in the last few years seems to suggest that perception and action are tightly linked rather than separated."[19] If so, we should expect mirroring wherever and whenever perception and physical or mental actions take place.

The brain mirrors what it sees and hears. That is how we navigate the world. Whether there are indeed cells specific to this task will remain unknown until we have unraveled the detailed neuroanatomy and physiology of every cell and every synapse, and their interrelationships—at best a wonderful dream.

Now let's take a moment to ask whether involuntary mental sensations might play a role in what appear to be excessive claims for mirror neurons. Ramachandran readily admits that "our current understanding of the brain approximates what we knew about chemistry in the 19th century."[20] But a quick look at his reasoning on mirror neurons brings us back to the familiar problem regarding our "sense of uniqueness." In the 2005 PBS documentary on mirror neurons, Ramachandran begins with: "Everybody's interested in this question: 'What makes humans unique?' What makes us different from the great apes, for example? You can say humor—we're the laughing biped—language certainly, okay? But another thing is culture. And a lot of culture comes from imitation, watching your teachers do something."[21]

Perhaps one of the major driving forces in modern neuroscience is the belief that we are unique and that this uniqueness can be established through biological evidence. How ironic, given that our own sense of uniqueness is itself driven by our biology. It is our sense of agency, ownership, and a unique sense of self that propels both our need to understand our uniqueness and the concurrent sense that we have the intellectual capabilities to make this determination. I am reminded of the myth of Sisyphus, where poor old Sisyphus is condemned for all eternity to push the rock up the hill, watch it roll down to the bottom, and then begin again. If it is our fate to have evolved a brain that believes it can solve a problem it is instrumental in creating, aren't we better off recognizing this paradoxical aspect of our biology rather than continuing to draw far-reaching metaphysical conclusions about the nature of man based upon our inherent mental limitations?

Perhaps even more ironic is that we would look to the presence of similar neural systems in monkeys and man to establish this difference, particularly since Ramachandran points out that, in his view, these very same mirror-neuron-equipped monkeys have no language, humor, or culture. Even if monkeys don't have well-developed language skills, what are we to make of the other modes of communication between other species? Is language the only form of communication that counts? If one of us speaks in English and another in sign language, we don't think of these as being fundamentally different as to purpose and function, but only as to form. As for other animals not having culture, you need look no further than Japanese snow monkeys (macaques), who have taught each other to enjoy sitting in hot springs, to make snowballs, and to wash potatoes instead of brushing off the dirt.[22] As regards humor, I have a friend with profoundly disabling parkinsonism. He's always had a wicked sense of humor and irony, but now, with his face frozen into an expressionless mask, he no longer exhibits any facial characteristics of being amused. There is no laughter or crinkling around the eyes, no grin or guffaw. He remains statue still. And yet, via a laptop, he can tap out, "LOL." If animals have a sense of humor but express it differently and don't have the capacity to tell us their feelings, we cannot conclude that they don't have a funny bone. Perhaps they are laughing on the inside, just like my friend.

It is hard to imagine how we might think of ourselves if we could step back from those involuntary mental sensations that steer our thoughts about the mind into a maze of blind alleys and inescapable paradoxes. Even so, such a vision must be our idealized albeit unobtainable goal. Though we cannot step outside the cognitive constraints imposed by these involuntary mental sensations, we can at least acknowledge the profound role they play in generating our thoughts about our minds. The mirror-neuron story should serve as a cautionary tale of good basic science being used to advance unwarranted claims about the unique nature of humans. If there is anything unique about the human condition, it is our biologically prompted feeling of uniqueness that drives much of contemporary thought about the human condition.

9 • Under the Big Top

What's he building in there?
What the hell is he building in there?

—Tom Waits, "What's He Building?"

On my sixtieth birthday, my wife and I took my mother to a particularly upscale San Francisco hotel for lunch. We wouldn't usually eat there, but the hotel had a convenient drive-up entrance and elevator for my mother to use her walker. Though we'd always been close, it would be fair to describe my relationship with my mother as undemonstrative. However, at the end of the meal, in an uncharacteristic gesture of affection and appreciation, I reached over, lightly touched her arm, and said, "If it weren't for you, I wouldn't be here today." Her immediate deadpan response: "We could have eaten somewhere else."

When she was ninety-seven and near death in the hospital, I mumbled an inanity meant to cheer her up. "You're lucky to have your room right across from the nursing station." Without any hint of making a joke, my mother replied, "Yes, location is everything."

Allegedly one of man's most unique attributes is his ability to read the mind of others (know what another is thinking). In philosophical circles this is referred to as theory of mind (TOM). But, even as a neurologist who reads philosophy and studies the mind, to this day, I have no idea if my mother was being humorous, sardonic, ironic, playful, insensitive, pragmatic, existentially whimsical, or matter-of-fact. Her deadpan delivery often made it impossible for her closest friends and family to distinguish between an intention to be funny or wry, or a genuine unawareness of possible double entrendres. In the end, each of us explained her Yogi Berra–like observations based upon his or her own unique lenses for seeing the world. As a result, she became the stuff of family legend, myth, and, ultimately, considerable confusion.

If we can't be sure someone is pulling our leg, can we accurately predict behavior? In a 1993 study, members of the Department of Psychiatry, University of Pittsburgh School of Medicine, evaluated patients originally seen in the emergency department of a metropolitan psychiatric hospital. When ready for discharge, the patients were assessed for their potential for violence and accordingly assigned to one of two groups—violent or nonviolent. In a six-month follow-up, violent acts occurred in half of the cases predicted to be violent, but also in over one-third of the "nonviolent" group. Within this group, predictions of female patients' violence were not better than chance. (Flipping a coin to decide violent or nonviolent had the same accuracy in prediction as did a psychiatric evaluation.) In another study, even such intuitively obvious differentiating factors as verbal threats versus actual prior physically assaultive behavior weren't shown to accurately predict subsequent violent behavior.[1]

Psychiatrists seem equally baffled with predicting suicide. A study at University of Iowa hospitals and clinics looked at 1,900 patients with major depression. The psychiatrists tried to predict those who would subsequently commit suicide; they failed to identify any of the subsequent suicide victims. The study concluded, "It is not possible to predict suicide, even among a high-risk group of inpatients."[2]

How about lying? Paul Ekman, psychologist at the University of California, San Francisco, and expert in micro facial expressions, prepared a videotape of interviews of ten men telling their views on capital punishment. Viewers were asked to decide which of the men was lying. Most people scored at chance levels or only slightly higher. Those you might think would be better than average—police officers, trial court judges, FBI and CIA agents—did little better than randomly selected bus drivers or pipe fitters. (A review of 120 similar studies revealed only two that reported lie-detection rates of 70 percent.) Explaining his reasoning for locking up Bernie Madoff for 150 years, the presiding judge, Dennis Chin, said, "Nine of Mr. Madoff's victims described the devastation he had caused in their lives. Mr. Madoff rose and offered a lengthy apology, saying he 'felt horrible guilt.' He turned to face the victims, and apologized again." Chin went on to describe Madoff as appearing sad, almost as if he were grieving. But in the end, Chin said, "I did not believe he was genuinely remorseful."[3] Most of us would agree with Chin, but it's not particularly heartening to know that we might well be wrong. What if there was a more accurate way to distinguish between Bernie putting on a good courtroom performance, having some belated remorse for others, feeling sorry for himself, or having some bad feelings but not knowing why?

Being relatively poor judges of character, it's no wonder that we look to science to help us out. Maybe neuroscience can find the neural correlates of various mental states contributing to behavior. In fact, a number of neuroscientists feel that mind reading is either here or just around the corner. Carnegie Mellon University neuroscientist Marcel Just is working on the use of fMRI in "thought identification," and told Leslie Stahl on *60 Minutes* that his team has already uncovered the "signatures" in our brains for kindness, hypocrisy, and love.[4]

John-Dylan Haynes of the Berlin School of Mind and Brain, Bernstein Center for Computational Neuroscience, goes a step further. He believes we will one day be able to determine people's intentions by looking at their brain activity.[5] Thomas Baumgartner of the University of Zurich envisions a future in which brain scanners might help psychiatrists decide whether or not to release on parole criminals who promise they won't break the law again.[6]

To see if any of these claims are even theoretically possible, let's sidestep discussion of the pros and cons of evolving technologies such as fMRI by projecting ourselves forward to a hypothetical time when we have the perfect brain recording device—let's call it a Superscanner—that can track every synapse, brain cell, electrical potential, and neurotransmitter at every instant. With the press of a button we can have a complete space-time map of brain activity. Would this allow us to accurately read another's mind? To understand his or her mental state?

To answer those questions, consider the first requirement of understanding a mental state. A basic principle common to all scientific studies is the establishment of a baseline against which physiological changes can be detected. Remember the old movies where a criminal suspect is given a lie detector test? To establish a baseline level of autonomic system responses to telling the truth or to lying, a subject is first asked a series of strictly factual questions such as his address, his date of birth, and the high school he attended. Readings are taken of various autonomic nervous system functions such as heart rate and skin sweating. By knowing how his nervous system responds when you have clear knowledge of whether he is lying or telling the truth, you have a measure with which to gauge his responses when asked more provocative questions such as "Did you kill Mrs. Jones?"

Though polygraph testing has been thoroughly discredited, the underlying principle of establishing a baseline response hasn't changed. No matter how advanced our measuring techniques become, we still need to know the baseline state of whatever we are studying. The perfect intracellular electrical recording will still require determining the normal pattern for cell firing in order to detect changes with the presentation of a new stimulus. With our Superscanner we will still need to obtain a picture of the "mind at rest," then present the subject with a mental or physical task and see what activity is present *above and beyond* the baseline measurement. Baseline measurements can be thought of as the equivalent of a control group in a randomized study. The difference is that for the individual subject, his baseline becomes his own control.

But baseline brain activity isn't synonymous with no brain activity. Our brain is performing all sorts of unconscious cognitive actions even when we are supposedly in idle. How are we to recognize and categorize cognitive activities that aren't reflected in a conscious mental state that we can accurately describe? Before addressing complex mental states, we should start by dissecting out the most elementary aspect of mind reading for which we do have objective standards—the prediction of a single motor action.

If you tap your finger, certain motor areas specific for finger tapping will light up on fMRI. Call this motor pattern A. You have knowledge of the input—your intention to tap your finger—and you can accurately record the output by measuring the speed, frequency, and strength of contraction of the appropriate muscle fibers. Having established this correlation in the individual, you can verify your findings by studying thousands of subjects. With this technique you can be confident that finger tapping will show up on fMRI as pattern A.

But the converse isn't necessarily true; seeing pattern A on the scan does not guarantee that you have tapped your finger. Case in point: mirror neurons will demonstrate the same pattern of activation with passive observation of an action and the intentional carrying out of the action. Detecting this pattern in isolation cannot tell us if the monkey is watching or making a grasping motion. We need to observe the monkey to make this determination. Those cells touted to be the basis of mind reading can't beat a coin flip in predicting observation or action.

Now up the ante. Rather than easy-to-quantify motor actions, try to understand unconscious cognitive activity. Imagine a future time when you, a neuroscientist, are using the Superscanner to study mental states in anesthetized patients. You discover a quite distinct and reproducible pattern of brain activity—pattern B—in some of the anesthetized (but not paralyzed) patients. You carry out detailed behavioral observations during surgery but see no difference in motor activity in those with and without pattern B. You conclude that pattern B doesn't represent a specific motor action, so it must be correlated with some mental state. Detailed interviews of the subjects as soon as they awaken from surgery are worthless; the subjects have no memories from their surgical experience. Personality profiles are equally unrevealing. You are convinced that you have discovered the neural correlate of a brain state, but you don't know what it is.

No matter how superb new imaging tools might become, neuroscience cannot explain pattern B without knowing exactly what each subject experienced at the moment pattern B occurred. We can only infer a correlation between brain state and mental state by relying on the subject's report of what he is experiencing. We all know how difficult it is to be fully aware of our mental states at all times, let alone adequately describe what we're feeling and thinking. Most of us acknowledge that psychological studies of mental states will always be less than perfect, as they can never completely overcome this inherent problem of subjective description. But assessing subliminal mental states runs into an additional impasse. If a subject is unaware of the state because his attention is focused elsewhere, because he has forgotten the moment and the feeling, or because he wasn't conscious during its expression, the neuroscientist has no response with which to correlate the pattern.

To see how this problem applies to unconscious mental activity, consider the challenge of determining intention—a major goal of mind reading. One of the roles of the conscious mind is to input intention into unconscious brain mechanisms. For example, you go to bed unable to remember the name of a popular song. In the morning the name pops into awareness. Your conscious intention to remember the name has been transferred to and carried out subliminally. Whatever unconscious cognition might be, we should all be able to agree that it is guided by intentions even when we are not consciously intending anything. Intention is often carried out without our being aware of it.

At any instant your brain is filled with myriad unconscious intentions—some short-term, others incredibly long-term. Right now you might be silently trying to remember where you put your passport, drumming up a new plot point that caused you to abandon work on a novel several years ago, and planning to work on your taxes over the weekend. Though we have no idea how such unconscious intention operates at a physiological level, it is likely that long-term intention is intimately connected to or embedded in a brain map that represents a problem to be solved or an action to be taken. If you are trying to decide where to go on vacation, the brain map might include all the various possible resorts and campgrounds, your personal likes and dislikes, and an embedded instruction for your brain to come up with the best solution. Our Superscanner can only record the presence of an unconsciously mediated intention in one of two ways—as part of the brain's baseline activity, or as a discrete neural pattern that stands out from the baseline. If the former, it won't be detectable. If the latter, we won't be able to recognize the pattern for what it is (we have neither objective nor descriptive access to unconscious intentions). Either way, unconscious intention is beyond the reach of neuroscientific inquiry.

I am tempted to draw an analogy between the study of basic brain mechanisms and cosmology. We see approximately 4 percent of what is "out there" in the universe. The majority of what we don't see—dark matter and dark energy—is defined by its effects on the visible universe, but is not directly detectable by conventional physics. Perhaps we should think of unconscious cognition in the same light (or lack of light). We can only know subliminal cognition by studying its effect on conscious states.

In science, a number of problems are resistant to empirical inquiry. We can get information that was generated incredibly close to the time of the big bang, but we cannot go all the way back to time zero; all understanding of the instant of the big bang is predicated upon inference from its subsequent effects. String theory can't be empirically validated, as the necessary measurements would require using an accelerator bigger than the earth itself. Similarly, unconscious cognition can be studied only by inference.

To put the problem of unconscious intention into a practical perspective, imagine the following scenario. Pete is on a vacation at a posh Caribbean beach resort. It's a glorious day; he's just finished an ocean swim, had a sumptuous buffet lunch poolside, and is now sipping on a large tropical drink. His mind is in neutral, devoid of any specific thoughts or even any vivid awareness of itself. At most, he is slightly aware of idly twirling the paper drink umbrella. If asked, he would say that "nothing is on his mind." A figure approaches. It is Mike, an old dorm-mate from his freshman year in college. He is much larger than Pete remembers; his belly is overflowing his more-than-ample Bermuda shorts. In quickly scanning his memory, Pete has the vague recollection that Mike was an okay guy; he makes it a point not to stare at Mike's midsection. Mike pulls up a chair, and they play catch up. Then, out of the blue, Pete suddenly remarks, "I see that you're still in great shape." No sooner are the words out of his mouth than his face flushes and he cringes with embarrassment. Mike gives Pete the finger and storms off.

Pete is puzzled. Why on earth would he make such a comment? Did he mean to insult him, or was it just idle banter gone wrong? Could it have been a Freudian slip, a reflection of unrecognized deeply rooted feelings about Mike? He retraces his mood to see what he can remember about Mike and similar situations when he's made a nasty or even modestly snide comment. Nothing makes sense. Further reflection doesn't help; Pete has no idea why he made the comment.

An additional piece of history, one that Pete doesn't remember: During the last weekend of their freshman year Mike and Pete went to a dance. Mike made a snide comment about Pete to a girl Pete had his eye on. Pete was humiliated but said nothing. Mike went off with the girl, and Pete was left licking his wounds. Pete hated being a coward, and vowed to get even with Mike. During the summer he fantasized a number of retribution scenarios, but upon returning to school in the fall, he found out that Mike had transferred to another college thousands of miles away. Over time, the slight receded from memory. Eventually it was fully forgotten. As far as Pete knew, he had put Mike out of his mind.

Not so for Pete's brain. He had repeatedly informed it of his desire to get even and had even given it some potential scenarios—his fantasy acts of retribution. His brain was just waiting for the opportunity to get even. But how can we know that this unconscious motivation prompted his comment if Pete can't tell us? Is there any method that will unlock this unconscious intention associated with a forgotten memory?

Until recently, improved self-awareness meant thinking about your psychological makeup and tendencies. Though many well-intended advocates of the dictum "The unexamined life isn't worth living" have spent years on the psychoanalytic couch, donned hair shirts, or roamed the self-help aisles of the local bookstore, the results have generally been less than impressive. The more science reveals the extent of unconscious brain mechanisms, the less we have come to believe in the value of psychological approaches to self-understanding. Long-term psychoanalysis and talking therapies have been largely replaced by behavioral modification, psychoactive drugs, and even brain stimulation. Encouraged by the ever-increasing number of neuroscientists who either explicitly or implicitly suggest that the precise identification of brain patterns can solve the problem, we are increasingly looking to science for self-understanding.

Let's put Pete to the acid test. We'll fit him out with our Superscanner (now available as a portable unit!) to wear poolside. You've got the perfect setup; his initial quasi reverie would clearly count as a neutral baseline. The sight of Mike approaching Pete would be easy to time and would serve as an excellent stimulus to mark the onset of any brain changes. When Mike enters the picture, all sorts of brain regions would fire up—from Pete's visual cortex constructing an image of Mike's former physique and contrasting it with his present state, to the frontal cortex deciding whether Mike might capsize the chaise longue if he sat down. Call this pattern C. Then there is the arrival of additional areas lighting up—the superimposition of a new pattern D on the already existing pattern C.

According to John-Dylan Haynes, expert in the rapidly expanding field of cognitive mind reading, "The new realization is that every thought is associated with a pattern of brain activity, and you can train a computer to recognize the pattern associated with a particular thought."[7] If this is true, we have pattern D as a potential starting point. All we need is to correlate this pattern with a conscious mental state. Unfortunately, Pete has no conscious awareness of any intention to insult Mike, and he has no memory of the inciting slight that triggered the long-term desire for retribution. The Superscanner has a pattern but no explanation. Your two options are to admit that you don't know the meaning of pattern D, or to try to make further correlations. When in doubt, collect more data.

You go on to examine a thousand subjects. Using standardized interview techniques and personality tests you find a perfect correspondence between pattern D and a passive-aggressive personality disorder. Your landmark study gives you instant celebrity status. After becoming an international expert on the snide comment, you are called to evaluate a young man who, for no apparent reason, suddenly fired a salvo of epithets at his boss. The man's job is at stake. The head of human resources will fire him if his insult was intentional but will give him a second chance if it's determined that he "didn't really mean it," as the man claims. Using your patented software, your scanner detects pattern D. The head of HR presses you and asks if this means the act is intentional. You answer that one of the hallmark features of passive-aggressive behavior is the intentional snide insult that the subject vehemently denies was meant to be insulting. (It is this discordance between denial of intention and our perception of intention that makes the passive-aggressive individual so utterly frustrating.) Based upon your assumption that pattern D has a perfect correlation with passive-aggressive behavior, and that insults from such folks are likely to be intentional irrespective of whether or not they admit their intention, the man is fired.

The transition from correlation with a brain pattern to a prediction based upon a psychology-based belief is seamless. You have accepted a psychological explanation of behavior—that passive-aggressive comments are intentional—in order to establish that pattern D implies intentionality. Such circular thinking is inevitable whenever studying a mental state in which intention is part of the definition of the state. Most of the major neuroscientific advances in the study of mental states have been in areas in which intention is not a critical feature. Fear is a perfect example. We can extensively study the various anatomic and physiological components of fear because we do not need to read minds and assign intention. The responses are spontaneous and reflexive—part of evolution's built-in circuitry to avoid danger. Similarly, we can study auditory and visual perception, both in animals and in humans, without needing to assign intention to the perceptions. On the other hand, it is impossible to conceive of hard data on altruism, generosity, compassion, truth telling, moral judgments, or other complex mental states without determining the degree of underlying intention.

Let me interject that I am not being critical of studies of human and animal behavior as a way of drawing commonsense assumptions about what our nature might be. Indeed, much of this book is predicated upon my use of such information to develop my personal view of a mind. But I am not offering my opinions as science. My concern is in using data drawn from behavioral observations as objective evidence for a state of mind rather than offering the data as detailed observations *as seen from an outside (subjective) perspective.* For all we know, pattern D could be nothing more than a marker for a genetic predisposition for increased irritability seen in those with passive-aggressive disorder. It could be a reflection of an underlying biological predisposition to Tourette's syndrome that isn't sufficient to create the full-blown clinical picture. If the latter, would you still call an outburst of epithets "intentional"?

Understanding intention is a primary goal of any serious attempt at mind reading. To know why X did Y, we need to understand X's intentions. If we want to know if X is telling the truth, we are also asking if X intended to tell the truth. If X says that he doesn't remember where he was the night that Mrs. Jones was shot, we can't, of course, directly know if he even tried to remember the night in question. Instead, we try to figure out whether or not his behavior suggests that he is trying to remember. Our seat-of-the-pants subjective assessment of intentions applies when we are judging an accused, deciding whether our president intended to carry out his campaign promises, or determining whether our teenage son is telling the truth when he says he is putting his maximum effort into his studies. Despite the wishes of neuroscientists, *there is no objective measurement of intention.* If you are a juror, your final decision as to the intent and therefore the degree of responsibility a defendant has will depend upon the story you tell yourself about the defendant's behavior.

The lack of a suitable animal model is another essential limitation on scientific attempts at mind reading. As animals can't tell us what they're feeling and thinking, we end up basing our opinions on what a particular behavior would reasonably mean if carried out by a human. If an animal gives a portion of his meal to a hungry comrade, we might see this as an act of sharing and evidence of empathy, generosity, and compassion. We read of altruistic whales and termites. For me, it seems entirely plausible that whales do contemplate the moral implications of their actions. It seems less likely that termites have the same capacity for compassion and long-term planning. But my interpretation is itself an example of using scientific data (the number of neurons in a whale versus in a termite and the different brain-size-to-body-size ratios) to justify what is pure speculation. No matter how precise the neuron counts, without interviewing each species, we cannot know the magic number that pushes an animal over the threshold into self-awareness, consciousness, or intention. Any attempt to justify my conclusions by pointing to brain size, architecture, and/or anatomy is the equivalent of the spindle-cell-neuron argument that we have just revealed as faulty thinking.

"Intention" is not a precise term. Neither is "purpose." Consider the slime mold. If we put a slime mold onto a topographical map of Great Britain, we can predict that the slime mold will consistently re-create an outline of the British highway system. At the very least, I think we would all agree that the seeking out of food is purposeful behavior. We would also agree that slime mold has no intention of specifically designing a highway system. But as we move up the complexity ladder, such decisions become increasingly difficult. Can we say that termites do not intend to build a termite mound because their brains are too small to carry the instructions, and presumably do not have conscious desires and intentions? If an individual termite never builds the smallest version of a termite mound, but always builds a mound in collaboration with his fellow termites, where does intention enter the picture? Once again we are back at the problem of how to objectively identify an intention at the neural level—either at the individual level or at the level of the group of termites.

Another way to think about this problem of different levels and manifestations of intention can be seen by comparing two medical conditions—a developmental disorder and drug addiction.

Lesch-Nyhan syndrome is a rare X-linked genetic defect characterized by a decreased level of, or the absence of, a specific enzyme, delayed motor development, moderate retardation, and characteristic self-mutilating behavior. If not suitably restrained, these children will spend much of their waking hours biting and chewing off their own lips and fingertips.[8] How would you classify such behavior in the presence of a clear genetic predisposition? Biting off one's fingers can't be considered a series of random acts if it is a behavior common to a particular genetic disorder. If not random, what should we call the behavior? Intentional? Purposeful? Involuntary? Purposeful but involuntary? Whatever we decide will depend upon how we define intention, purpose, and volition, and the degree to which we restrict these labels to conscious states of mind. What if the child knows that he will be punished (restrained more securely) but "cannot help himself"? (Some say that the explosions of foul language sometimes seen in Tourette's syndrome fall into this general category: intentional but carried out in order to satisfy an uncontrollable urge.) I suspect that most of us would opt for some intermediate interpretation: that the behavior is both involuntary and intentional in the sense that the brain is sending out specific motor impulses with the intention to bite the fingers.

Compare this situation with drug addiction. Though science has shown that our reward systems literally crave the missing drug, we hold the addict at least partially responsible for his behavior. The implication is that his intention to continue taking drugs is under some degree of self-control. In this case we assign a degree of intention based upon our larger view/wish/understanding/hope regarding how we can/should/do behave. It is hard to imagine assigning intention without incorporating one's personal moral perspective. Similarly, how we view intention will always be influenced in part by our own experience of and understanding of the sense of agency. If there were a designer of the human condition, this would surely count as delicious mischief—creating involuntary mental sensations that contribute to our deciding how much voluntary intention is involved in a thought or an action, and then having us use this interpretation to attempt to create a fair system of social order.

There are further arguments but the point seems pretty self-evident. If intention can exist outside of conscious awareness and cannot be directly studied, understanding of any mental state that requires an assessment of intention must remain incomplete.

Rush to Judgment

A bit of neurological history: Before scanning techniques were available, electroencephalography (EEG) was the primary tool for correlating brain activity with behavior. Certain

EEG patterns were correlated with a variety of mental states and disorders—schizophrenia, depression, obsessive-compulsive disorder, and certain personality types, just to name a few. The decisions based on these correlations were often catastrophic. One of the more tragic and avoidable outcomes of overinterpreting the significance of these EEG patterns occurred in those criminals with recurrent episodes of violent behavior. Correlation was confused with causation. The assumption was that the EEG patterns reflected a basic brain instability that directly produced attacks of violence in the same way that a short circuit could trigger a seizure. If true, the thinking went, then controlling or removing the affected area of the brain should prevent the outbursts.

As a result, chronic offenders were subjected to radical treatments ranging from attempts to medically suppress the EEG pattern to lobotomies to electrode implantation and direct brain stimulation. When none of the treatments worked and long-term follow-up studies failed to accurately predict further violent episodes, the original EEG findings were reinterpreted. Now the patterns were felt to represent prior brain damage common to those with a history of physical abuse or frequent fights. Cutting out the involved areas of the brain no longer made physiological sense. A generation's worth of well-intended but ultimately unnecessary and often harmful treatments could have been avoided if neuroscientists had carefully considered possible limits to the role of EEG in determining causes of behavior. Sadly, this pattern of initial wild enthusiasm for a particular neuroscience finding followed by a more sober reconsideration of other possible interpretations isn't going to go away. The negative lessons of history are far less seductive than the positive findings offered up by a new scientific technique.

Mind reading is a prime example. In 2008, based upon a proprietary quantitative EEG technique (BEOS or Brain Electrical Oscillations Signature), a twenty-five-year-old Mumbai MBA student was convicted of murdering her ex-fiancé by serving him candy laced with arsenic. The circumstantial evidence was slight to nonexistent; the accused vehemently denied any involvement in the crime. But she agreed to the BEOS test in order to establish her innocence. The test indicated the woman had undeniable "experiential knowledge" of the crime. The judge felt that the proof of the case hinged on the science; she was convicted of murder and jailed for life.[9]

Over the next six months the BEOS test provided evidence instrumental in the conviction of two others accused of murder. Meanwhile, in September 2008, a report by a committee at India's National Institute of Mental Health and Neurosciences (NIMHNS) declared that such brain scans of criminal suspects were unscientific. The former MBA student used this finding to appeal and is now out on bail; because of the slowness of the Indian judicial system, she may wait five to ten years before her appeal is heard. When hearing of the case, Geraint Rees, professor of cognitive neurology at University College London, said, "There is nothing in the history of brain imaging to say that we could ever get the degree of precision needed to detect lies." Nevertheless, India's chief forensic scientist continues to claim, "The technique has great potentiality to become an infallible tool in crime investigation. It can become a revolutionary technique like DNA fingerprinting if its evidential strength and judicial acceptability are established."[10]

Closer to home, we have a number of companies offering imaging-based lie detectors. One, Cephos, claims that its fMRI-based lie detection has tested more than three hundred people with an accuracy ranging between 78 and 97 percent. Even if true, this would mean an error rate as high as one in five. More critically, there is no reasonable methodology to establish a false-positive rate. If you get a positive mammographic reading suggesting breast cancer, you can perform a biopsy. A false positive can be determined by looking at the biopsy specimen and not seeing any tumor cells. But what technique do you have for establishing a false-positive fMRI reading? There isn't any independent test for establishing the truth of mental states.

Even so, the CEO of Cephos, Steven Laken, Ph.D., claims that the technology can be developed into a new discipline of forensic brain scanning that examines people's intentions, motivations, and feelings. "Does someone understand that what they did was wrong, or did they intend to do it? This makes the difference between murder and manslaughter. We may also be able to tell if someone has been in a terrorist camp, or had certain motivations. For example, if you show someone a place they recognize, their brain reacts differently under fMRI than if they are seeing a picture of a place they never visited. With eyewitnesses, false memories light up different parts of the brain than true memories, which could be very useful for asking witnesses to identify criminals." Even more astounding is Laken's claim that with the proper commitment of resources, this forensic technique for lie detection and judgment of motivation could be fully developed in about a year.[11]

To reiterate: rather than continue to believe that a new or improved technology will provide the necessary information to demonstrate a subject's intention, both consciously and subconsciously, we should acknowledge an essential limit of neuroscience: intention is not a mental state that can be captured via any known scientific test. So whenever you see the claim that a technique can unravel intention and motivation, run, do not walk, to the nearest exit.

10 • Consciousness Unexplained

Whereof one cannot speak, thereof one must pass over in silence.

—Ludwig Wittgenstein, Tractatus Logico-Philosophicus

Most of us either have had or will have the dreadful experience of being at the bedside of a close friend or family member rendered unconscious by a major car accident, heart attack, or stroke. There are few if any clinical situations more desperate than trying to imagine what is going on inside an unresponsive patient's mind. Is he in pain? Does he know what has happened to him? Does he want every conceivable treatment or to be put out of his misery? Being unable to communicate with the patient, we are left trying to imagine how we would feel in a similar circumstance.

Until recently, in these kinds of situations, we have been dependent upon the clinical neurological assessment to determine a patient's level of consciousness, as well as the likelihood and degree of potential recovery. Though neurologists have developed a number of behavioral guidelines that are relatively good predictors of long-term prognosis, these are far from foolproof. In addition, our best bedside observations shed little light on the victim's state of mind, whether he is aware of his condition, and if so what course of action he would prefer. We can't even learn retrospectively. Patients with a severely altered level of consciousness almost invariably have profound memory disturbances; those who do recover are unable to accurately describe the experience. Any improvement in our understanding of the mental state of the unresponsive patient would be a huge advance.

Given the limits of bedside neurological evaluations, it is only natural for neuroscientists to turn to modern technology for better clarification of a nonresponsive patient's mental state. But is this realistic? Can sophisticated imaging techniques improve our understanding of the mental life of a clinically unconscious patient? As you read about the following case, see if you can determine whether the patient was conscious and aware of her environment, and if so, to what degree? But before we get started, here are some quick definitions of the formal levels of consciousness.[1]

Coma: The complete failure of the arousal system with no spontaneous eye opening or ability to be awakened by application of vigorous sensory stimulation.

Persistent vegetative state (PVS): The complete absence of behavioral evidence for an awareness of self or the environment. There is preserved capacity for spontaneous or stimulus-induced arousal, evidenced by sleep-wake cycles. "Persistent" refers to being present for at least one month following the brain insult.[2]

Minimally conscious state (MCS): Inconsistent but clearly discernible behavioral evidence of consciousness such as following simple commands, making gestures or verbal yes/no responses (regardless of accuracy), intelligible verbalization, and/or purposeful behavior.

Locked-in syndrome: The inability to speak or move the arms and legs, but preservation of awareness and cognitive abilities. Mainly occurs as the result of an injury to the upper brain stem that spares higher cortical functions and is not a disorder of consciousness.

A twenty-three-year-old woman, patient X, was involved in a major car crash. She was brought to the hospital in a coma. Scans revealed multiple areas of severe brain damage (areas of bruising) as well as diffuse brain swelling and extensive bleeding within the brain. Scans also showed some remaining intact areas of cerebral cortex.

The woman underwent several neurosurgical procedures but did not improve. After six months she was able to open her eyes spontaneously. She did move her arms and legs reflexively to pain, loud noises, and unpleasant odors, but did not make voluntary movements on command, nor was she able to follow a moving object with her eyes. When asked questions, she neither responded verbally nor maintained eye contact for more than a few seconds. She had a normal wake-sleep cycle. After six months, a number of consulting neurologists agreed that the patient had been left in a persistent vegetative state (PVS).[3]

To assess her level of consciousness, a research team from England and Belgium headed by neuroscientist Adrian Owen came up with an ingenious use of the fMRI. The unresponsive patient was placed inside the scanner and asked to perform two separate mental tasks: "Imagine playing tennis" and "Imagine visiting the rooms in your home." In order to generate a rich and detailed mental picture during the imaging, the patient was asked to envision playing a vigorous game of tennis using both forehand and backhand for the entire duration of each scan. Similarly, she was asked to visualize slowly moving from room to room in her house, noting the location and appearance of each item of furniture as she did so.

When the patient was asked to imagine playing tennis,

increased brain activity was observed in the motor area of the brain that lights up when normal volunteers imagine playing tennis. When asked to walk around her house, she demonstrated increased brain activity in regions normally activated during real or imagined spatial navigation. Despite being clinically unresponsive to her environment, her fMRI responses on these two tasks were similar to those of fully conscious and actively cooperating volunteers. The researchers' conclusion: "This patient was consciously aware and willfully following the instructions given to her, despite her diagnosis of being in a vegetative state."[4]

For me, this study raises a number of questions central to present-day neuroscience. How can we distinguish between conscious and unconscious cognition? Can seemingly willful acts be carried out entirely outside of consciousness? Does the carrying out of a "willful" mental act necessarily indicate that you are self-aware? The operating assumption of Owen and colleagues is that the patient's ability to imagine herself playing tennis or walking around her house is prima facie evidence of willful conscious behavior. This presumes that without consciousness these two tasks could not have been performed. But is this the only possible explanation for the similarities in fMRI patterns, or is there a possible alternative explanation that does not require any conscious awareness or intention on the part of the patient? To address this problem, let's look at a combination of real-life experiences, thought experiments, and some provocative brain studies.

At a cocktail party, you are intently talking to one person and have tuned out other conversations. Suddenly you are aware that someone across the room has mentioned your name. The conscious "you" did not will this action at this particular time; rather than consciously eavesdrop on all the other conversations, you have long ago learned to assign this task to unconscious brain processes. An auditory input—the sound of your name—triggers the subliminal act of recognition. This same argument applies to any incoming sensory stimulus, be it a face in a crowd or the fragrance of a flower.

What we see or hear or how we experience pain or love is not a conscious determination. What consciousness does provide is the ability to focus our attention on incoming sensations; consciousness lets us pick and choose what to look at or listen to. But we cannot directly affect the perception of the experience. For example, if you stub your toe, you can try to think about something else, but you cannot consciously change the feeling of pain into a frisson of joy.

This is old news. None of us are surprised that sensory inputs are processed unconsciously. Conversely, it is hard to imagine the utter chaos if this weren't true. I am reminded of the era of telephone party lines when multiple phone conversations could be heard on a single line. Just one other voice in the background made conversation difficult to impossible. Now, up the ante from two voices on a line to the enormous number of sensory inputs that your brain is constantly processing. It makes perfect evolutionary sense that we would have developed brain mechanisms for processing sensory inputs outside of consciousness and then making us aware of only those inputs that demand our conscious attention.

If sensory inputs routinely trigger unconscious cognitive activities, we need to establish the upper limits (if any) of this automatic processing before we can begin to understand the role of conscious thought. Perhaps the easiest way to sort through studies on conscious thought is to constantly ask if a particular mental state could have occurred exclusively as a result of unconscious brain activity.

If the answer is always yes, then you are an epiphenomenologist—a believer that the conscious mental state contributes nothing to our overall cognition. Short of this position, each of us will have his own take on the overall role of conscious thought. This view will be in part determined by his individual sense of mental agency and his subjective "sense of control" over his thoughts (the inescapable problem of involuntary mental sensations dictating how we conceptualize and investigate the nature of our minds).

Determining the mental state of patient X must begin with an understanding of how we produce mental imagery. As an example, think of your home. Without prior exposure and a working knowledge of the layout, it would be impossible to willfully generate a mental floor plan. When you move into a new house, you learn all the particulars—the distance from the door to the bed, the proximity of the night table—that eventually allow you to navigate in near darkness. This knowledge is stored as a representational map labeled "my home" and will, on request, produce a mental image of your house. Such mental imagery can be elicited with direct brain stimulation studies—if the appropriate part of your brain is stimulated, you will "see" the layout of your home. The mental map can also be seen in altered states of consciousness; in our dreams we commonly walk around our house. (And if you are a tennis player, like patient X, you might envision a tough match while sound asleep.) But such observations beg the question of whether a verbal request to generate mental imagery must be consciously processed (consciously understood and acted upon) or can trigger the activation of the mental imagery without any conscious intervention. Though no single line of reasoning can give us a conclusive answer, we can sneak up on this question and get some partial answers, as well as an understanding of the limits of this question.

Your beloved spouse, sound asleep on his back, is making a horrible racket with his snoring. You lean over and ask him to roll over. He obliges, though his eyes remain closed, his breathing deep; there is no sign of him awakening. In the morning he has no recollection of the event. How are we to interpret this behavior? Probably most of us would offer some intermediate interpretation—that he awakened just enough to hear your request and follow the instruction, but not enough to have an overt change in behavior (apart from rolling over) or to store the event in his memory. But this explanation is pure guesswork, and lumps together two quite separate mental processes—arousal and awareness.

Many years ago, after playing in a nearly all-night poker game, I was too tired to drive back home over a fifty-mile winding road. I phoned my wife, checked into a local Howard Johnson and promptly fell into a deep asleep. About an hour later my wife phoned back to remind me of a meeting later that day. I can still recall the feeling of being startled by the sound of the ringing phone, jumping out of bed in the unfamiliar darkened room. I was wide awake but with no idea where or who I was. Though the details of the event are now long gone, what I remember most distinctly was the sense of utter alarm and confusion. The moment seemed to last an eternity although in real time it was probably at most a second or two before my sense of self and understanding of where I was reappeared. This transient disconnect between being awake and being aware of myself and my surroundings remains a vivid memory. That night I realized for the first time that a sense of self was not an invariable accompaniment of consciousness and instead must represent a separate physiologic process from "being conscious." It was as though my sense of self had to be projected onto a mental state of consciousness in the same way that a motion picture is projected onto a blank screen. Many of you will have seen this discordance in family members or friends with severe dementia.

Full consciousness remains even as the sense of self slowly dissolves.

Patients with dementia also demonstrate the striking discrepancy between motor memory/motor skills and self-awareness. When Ronald Reagan had moderately advanced mental deterioration due to Alzheimer's disease, he could still play a decent game of golf. We've all seen dementia patients who have normal motor skills even when they don't recognize their family or friends or even know who they are. Some can sing songs even though they've fully lost the ability to communicate with speech. They are able to feed themselves long after they have lost the ability to recognize what they are eating. The standard explanation is that implicit motor memories—the representational maps for eating, playing golf, or singing—remain relatively intact even though the higher cortical mental functions are not. Mental motor actions such as imagining playing tennis or walking around the house aren't equivalent to the demonstration of higher-level cognitive abilities. FMRI activation of brain regions for initiating motor actions isn't synonymous with a conscious self that is aware of this action any more than absentmindedly scratching an itch indicates a conscious intention and awareness of scratching.[5]

Similarly, when speaking of consciousness, it is important to distinguish between arousal/attention and actual self-awareness. With your sleeping spouse, even if he were to fully awaken as a result of your whispering in his ear, this does not mean that he would be self-aware of his actions any more than I was aware of where and who I was just because I was standing alongside the bed fully alert answering the phone. Full arousal tells us nothing about a person's self-awareness at that moment. Knowing that a person "is conscious" cannot provide us with any insight into the "contents of consciousness." I cannot imagine a more frightening scenario than having that sickening moment of disorientation and loss of sense of self in the Howard Johnson persist for any length of time. Yet from an outsider's perspective, my behavior would look normal and presumably willful—getting out of bed to answer the phone. My private nightmare wouldn't be detectable unless I could describe it. But such a description would require that I emerge from the nightmare with a full recollection of how I had felt at that moment. As even a minor concussion impairs our ability to store and recall memories, we can never know the quality of the mental states of someone with vegetative state, with minimally conscious state, or indeed, suffering from any significant alteration of consciousness.

In addition to arousal not equaling awareness, we have the problem of "being conscious" not being synonymous with consciously willing an act. We've all had the experience of driving on autopilot. If, during such a drive, someone says to you, "Turn left," you may well do so without any conscious awareness of hearing or carrying out the instruction. Someone screams, "Brake!" and you jam on the brakes before you are aware of any intention to do so and before you see the dog running across the road. The incoming auditory stimulus directly provokes a motor action without requiring any conscious intervention. If the hypothetical neural correlate of consciousness were to be discovered, it would show that the driver was clearly conscious at the time that he jammed on the brakes. But it couldn't tell us whether jamming on the brakes or any other specific action was consciously determined. Being conscious and a conscious determination of a particular thought or action are two interrelated but independent brain functions. As previously mentioned, determination of conscious intention must begin with an understanding of unconscious intention—about which we presently know nothing.

Let's try another approach. With the advent of functional brain imaging, numerous studies have documented residual and presumably automatic brain processing in patients in a vegetative state. In 1997 it was shown that the speech-processing areas of a PVS patient consistently responded when his mother read him a story.[6] Other studies showed patients' speech-recognition regions responding to hearing their own name but not the names of others. The degree of activation inversely correlates with the depth of unresponsiveness. With the more profoundly injured, these responses will be limited to primary speech areas; with lesser degrees of injury, higher-level cortical association areas (regions that coordinate and interpret more basic brain activities and shape them into our perceptions) can also be activated. This higher-level but still subconscious sensory processing is limited to those patients with multifocal areas of injury who also retain some degree of residual neural function in other regions of the cerebral cortex. Those with PVS as the result of a diffuse brain insult (i.e., lack of oxygen from a cardiac arrest) rarely show evidence of such processing. The same automatic processing occurs with visual inputs. For example, a PVS patient (whose eyes were open but who was unable to move on command) has been shown to have visual areas for facial recognition light up in response to familiar faces but not to unfamiliar faces.[7]

These fragments of presumed cognitive activity most likely are a reflection of the modular nature of our brains.[8] Uninjured areas of the brain continue to produce isolated aspects of a perception even when higher-level brain processes are too damaged to integrate and complete the perception and send it into consciousness. For example, to understand speech, higher-level brain association areas gather together lower-level

inputs such as the processing of words, syntax, context, nuance, and body gesture to give us an overall sense of the meaning of the simplest sentence. If an area for recognizing a word
escapes injury, it will continue to respond to an incoming auditory stimulus even though there may not be any conscious
recognition or understanding of what was said.

The key point: as long as areas for the processing of various
aspects of sensory input remain functionally intact, they will
continue to do their job irrespective of the patient's overall
state of consciousness. With fMRI studies of patients in various degrees of altered consciousness, we are getting a progressively more detailed picture of the functional anatomy and
hierarchical structure of unconscious cognition. Lower-level
functions contribute to higher-level cortical functions until, at
some point, they morph into conscious experiences. How this
transition to conscious experience occurs is anyone's guess.

To get a sense of the magnitude of the problem of studying
this transition from detectable neural activity to conscious
mental states—the hard problem of consciousness—consider
the following example. A hundred billion people are standing
in front of a huge lamp that turns on and off via an electrical
switch controlled by a single giant lever. It takes the combined
effort of all one hundred billion people to pull the lever. On
command, all exert full effort. One man with a torn forearm
muscle puts out maximal effort but isn't able to generate any
significant force. This slight difference is enough to prevent
the lever from being pulled. The lamp isn't switched on.

Meanwhile, a similar number of people perform the exact
same task with everyone exerting full effort, and the light comes
on. If we were comparing the activity of the two groups, we
would see no difference in the amount of effort exerted, either
individually or collectively, even though one group would
turn on the light, while the other wouldn't.

Now make these people neurons. If we had intracellular electrodes registering the firing of every neuron, we would still see no difference. Every neuron, including the one activating the torn muscle, fired normally. As the fMRI measures the metabolic demands of the brain, equal activity of all neurons in both groups would create identical scans. But here's the rub: these identical images would be associated with different outcomes. The failure to switch on the lamp isn't at the level of the neurons, but at the level of the injured muscle being unable to exert normal force. The fMRI would not be able to predict whether or not the light (of consciousness) was actually turned on.

This problem of levels cannot be overstated. It is central to the limits of neuroscience. Just as looking at quarks won't tell us anything about the properties of a carbon atom, and examining carbon atoms won't reveal properties that we associate with "life," lower brain states cannot reveal properties that emerge at a higher level of complexity. Consciousness isn't in the neurons any more than group behavior is present in the individual slime mold or locust. To believe that we can find the neural correlate of consciousness is to believe that looking at neurons and their connections is sufficient to characterize behavior organized at a higher level of complexity. To me, this represents a mistake of categories that cannot be overcome with better machinery and techniques. Only when we can bridge the gap between basic physiology and higher-level emergent properties, and express this understanding in physiological terms, will we have any theoretical chance of a neural signature for consciousness.

Hopefully the preceding discussion will have highlighted some of the theoretical problems in determining the mental state of patient X. But there are also huge moral questions. If you were her consulting neurologist, how would these fMRI studies shape your prognosis and treatment? What can or should you tell the patient's family? If you include patient X in a study you are conducting, how will the wording of your findings affect other PVS and MCS patients and their families? To address these questions, let me provide some practical context.

It is conservatively estimated that in the U.S. alone there are approximately 35,000 people lingering indefinitely in a persistent vegetative state, another 280,000 in a minimally conscious state.[9] Not even taking into consideration the patients who remain under the radar, being cared for at home by their families, the annual cost of medical treatment runs in the billions. The prognosis gets progressively worse the longer a patient remains in either VS or MCS. Few patients who are in PVS for more than a few months recover to full independence; most remain severely disabled.[10] To date, there is no convincing evidence that rehabilitation efforts significantly increase the chances for independent living. How imaging studies on clinically unresponsive patients are presented to the general public will have a profound impact on hundreds of thousands of patients and their families.

The study of patient X has yielded some dramatic and potentially practical results. In 2010, a twenty-nine-year-old noncommunicative auto accident victim with MCS was able to willfully modulate his fMRI.[11] The researchers were able to train the man to answer simple yes/no questions by either imagining playing tennis or walking around the house. Activating the "playing tennis" circuitry was taken as a yes, the "walking around the house" circuitry as a no. As these two brain areas are relatively far apart, it was easy to distinguish between the two responses. The researchers were able to establish that the man was able to answer five out of six questions correctly. Questions were of the order of complexity of

"Do you have any brothers?" and "Is your father's name Alexander?"[12]

These are impressive findings. If verified, they speak to the likelihood that conscious patients with impaired communication will finally be able to be heard. But this technique won't be able to determine the degree of consciousness and its contents unless the patients can (via the technology) adequately describe their inner mental states.[13] Also, it is folly to think of consciousness as a binary condition—a single mental state that is either present or absent. Consciousness includes a variety of mental states, ranging from fully alert and oriented, to the amnesia and confusion seen with a concussion, to the delirium of high fever and drug intoxication, to the dissociation of being awake yet without self-awareness (as in my awful Howard Johnson moment), to the sheer terror of an interminable nightmare. A primary goal in studying behaviorally unconscious patients is to get a better idea of the totality of their mental life—their mental capabilities and their fears, joys, and desires—not simply whether or not they are conscious. To this end, future technological advances will continue to produce better transcription devices—mental typewriters that will jot down what a person is trying to say. If we want to know their inner experiences, we will still need to ask them directly.

Since the study of patient X appeared in 2007, there has been a flurry of criticisms about the use of the fMRI confirming the presence of consciousness, as well as potential ethical conflicts. Here are three well-considered responses:

Nicholas Schiff, a Cornell neurologist specializing in the study of altered states of consciousness and a sometime collaborator with Owen on studies of PVS: "It is the case that we cannot confirm awareness simply on the basis of imaging findings without some reliable communication from the patient. . . . The selective identification of cerebral activity alone may not demonstrate the recovery of cognitive function.[14]

Lionel Naccache, a French neurologist studying consciousness: "In the case of Owen's vegetative patient who imagined playing tennis, it's impossible to know whether she reported the event to herself—which would suggest that she is capable of conscious thought—or whether . . . she had no subjective awareness of the experience. . . . Brain-scan research cannot yet tell us much about such a patient's prospects for improvement."[15]

The *New England Journal of Medicine*: "Without judging the quality of any person's inner life, we cannot be certain whether we are interacting with a sentient, much less a competent, person."[16]

To respond to criticisms, Owen and colleagues wrote: "Of course, alternative explanations will always remain theoretically possible—likewise, it is theoretically possible that none of us are consciously aware and behavioral responses throughout life are merely the result of 'automatically' triggered activity in our brains."[17] In 2006, in the journal *Science*, Owen argued that patient X's "decision to cooperate with us by imagining particular tasks when asked to do so represented a clear act of intent that confirmed beyond any doubt that she was consciously aware of herself and her surroundings, and was willfully following instructions given to her, despite her diagnosis of a vegetative state."[18]

I have presented verbatim the two sides of this debate in order to point out a central difficulty of the translation of neuroscience data into psychological explanations. I cannot imagine a more succinct demonstration of the insistence upon an absolute state of knowing despite considerable contrary interpretations than Owen's conclusion to his evaluation of patient X. This difference between an open-minded scientific approach and a more restrictive and dismissive personal vision couched as science can be seen in the language of the various authors' conclusions. Contrast the critics' "may be an example of," "may be nothing more than," and "it's impossible to know whether" with Owen's "clear act of intent" and "confirmed beyond any doubt." The former is the language of science exploring all possibilities; the latter is the language of agenda.

Earlier I discussed how our logic is influenced by a sense of beauty and symmetry. Even the elegance of brain imaging can greatly shape our sense of what is correct. In a series of experiments by psychologists David McCabe and Alan Castel, it was shown that "presenting brain images with an article summarizing cognitive neuroscience research resulted in higher ratings of scientific reasoning for arguments made in those articles, as compared to other articles that did not contain similar images. These data lend support to the notion that part of the fascination and credibility of brain imaging research lies in the persuasive power of the actual brain images." The authors' conclusion: "Brain images are influential because they provide a physical basis for abstract cognitive processes, appealing to people's affinity for reductionistic explanations of cognitive phenomena."[19]

If we accept that our assessment of our reasoning is itself shaped by involuntary elements, it follows that we should take great caution in issuing black-and-white conclusions to cognitive studies relying on a particular line of reasoning. In the end, without the ability to communicate with patient X directly, our conclusion as to whether or not she is conscious is based solely upon the line of reasoning we believe is most correct. Meanwhile, millions of lives will be directly affected by Owen and colleagues' impossible-to-substantiate assertion that she is unequivocally awake and aware.

Imagine a close family member as a victim of a near-fatal traffic accident. She survives but remains in a PVS for a year. Before the accident, she had written a living will requesting that all nutrition and support should be withdrawn if she was ever to end up in a vegetative state. She has discussed these wishes with you, her closest family member, in detail, and you have agreed to abide by them. After a year, the neurologist tells you that further recovery is quite unlikely. You spend some time at her bedside trying to imagine what she might or might not be experiencing. Eventually you reassure yourself that she isn't aware of her situation and advise her physicians to withdraw her care. Afterward, you still wonder if you've done the right thing, but tell yourself that you were acting according to her wishes and your promise. How will you feel when you subsequently read that a patient in a similar condition is unequivocally conscious and aware of her surroundings?

Friends and families frequently grasp at straws when faced with a medical disaster; they are likely to embrace the most improbable prognoses or therapies or false hopes, often at considerable emotional and financial expense. When presenting the results of the study to the family of patient X, wouldn't it be preferable to say that there are some inherent limitations to the study that make it impossible to say whether or not patient X is conscious and aware? The underlying caveat, applicable to all clinical neuroscience studies: The urge to offer conclusive opinions on new and/or controversial research must take a back seat to the higher priority to, above all, *do no harm*.

11 • Anatomy of a Thought

> Somewhere between the lines, the great truth is hidden. It has rolled under the sofa, and if we could only extend our fingertips a little bit farther, we would have it in our hands.
>
> —Vladimir Nabokov

Shortly before I finished my neurology training, Dr. A, the chief resident in internal medicine and later chairman of a major medical school department, asked me to join him in a long-term research project. I asked him what he had in mind.

"We can study alcoholism."

"What specifically would you be looking at?"

"Blood, urine, spinal fluid—whatever we can get. With enough samples, we're sure to find some previously undiscovered abnormality. It's a no-brainer."

"But what are you looking for?" I asked again.

Dr. A shrugged. "We'll know when we see it."

One of the basic premises of the biological sciences is that a detailed understanding of the anatomy and physiology of a system is synonymous with understanding the function of that system. If you know the contractile force of a muscle fiber and the number and type of muscle fibers in a particular muscle group, you can quickly calculate the amount of force that muscle can exert. If you go to the gym and bulk up, you can confidently predict that an enlarged biceps muscle will be able to exert greater force than prior to weight training. By knowing the relevant composition of your muscle fibers—the percentage of fast-twitch to slow-twitch fibers—you can make a quite accurate prediction as to whether a person will be better at sprinting or running a marathon. The equation common to all branches of the biological sciences: anatomy + physiology = function.

But applying this equation across the Great Divide—from the brain to the mind—doesn't necessarily work. Trying to make scientific observations that work at the level of the objective brain and also explain actions at the level of the subjective mind has resulted in some stunning neuroscience confusions, misrepresentations, and unwarranted fantasies.

Einstein's Brain

From a 2010 *New Scientist* magazine: "Oversized brains are to humans what trunks are to elephants and elaborate tail feathers are to peacocks—our defining glory. What would we be without our superlative, gargantuan, neuron-packed brains?"[1]

Is this true? Does bigger mean better? Though few of us believe that genius is simply a matter of anatomy, a great deal of time has been spent and folklore generated in trying to find out if Einstein had some anatomic variant that could explain his brilliance. When Einstein's brain's weight and size turned out to be unremarkable, more fine-grained explanations were sought. Glial cell advocate Andrew Koob has said, "Einstein's genius stemmed from an abundance of astrocytes populating brain areas involved in math and language."[2] But as the norm for the absolute number of glial cells isn't well established,[3] it's hard to say what "an abundance of astrocytes" means. Besides, increased astrocytes can be due to other causes, such as prior trauma leading to brain scarring. If we go down the absolute-numbers path, it has been said that Einstein had a 15 percent larger inferior parietal region, the part of the brain associated with mathematical thought and the ability to visualize movement in space. But how should we translate a 15 percent increase in the brain region? It has even been suggested that Einstein's brain lacked a particular groove (sulcus) that normally runs through this parietal region of the brain, allowing the neurons on either side of this missing groove to communicate more easily.[4]

Looking at brain size and shape as measures of the man is nothing new. The most famous early example—phrenology—exploited our tendency to make correlations whenever possible. Its original proponent, the Viennese physician Franz Joseph Gall, claimed that he could determine various traits of character and intelligence by examining the shape of a subject's head. During its heyday, some extraordinary claims were made. A number of European phrenologists argued that smart people have big brains and that other races were dumber because of their supposedly smaller heads (based upon their selective use of head measurements).[5] Doctors appeared as expert witnesses in court cases, using phrenology to explain the defendant's character. Others used "bumpology" as a form of psychoanalysis. (I confess to having an original phrenology head in my office—given to me by a former Oxford professor of neurosurgery. When puzzling over an idea, I sometimes feel a bump near the top of my head that phrenologists thought represented the site for speculation. And I have an indentation where the causality center should have been.) Despite extensive criticism and a thorough debunking as far back as 1829, phrenology remained intermittently in fashion throughout much of the nineteenth century, finally disappearing at about the same time as Freudian psychoanalysis came into vogue.

We now deride phrenology as an icon of pseudoscience.

But phrenology does offer us some deeper lessons. The originator, Dr. Gall, is also credited with being one of the first modern-day proponents of correlating brain regions with specific functions.[6] His underlying premise—that different areas of the brain perform different tasks—remains a central tenet of modern neuroscience.

The basic problem of phrenology and modern neuroscience, though, is knowing whether a particular technique is actually measuring what you intend it to measure. To get a sense for how our perception skews our understanding of science, take a look at an Oxford professor's 1859 assessment of phrenology:

> Calmly viewed, phrenology exhibits only a set of the most unexpected relations, at first collected and examined in the most purely empirical manner, in complete absence of any theory; out of which, by slow degrees, a system has been elicited, of which it can only be said, that at present it exhibits just that sort of rough, general coherency which, in spite of numberless objections in detail, gives an assurance of something too deeply seated in truth to be put down as mere random coincidence or fanciful delusion.[7]

The argument: Blind data collection produces general patterns and correlations that can be counted on to represent a deeper truth, irrespective of conflicting details. This is the equivalent of Dr. A suggesting that we collect blood, urine, and spinal fluid specimens and see what turned up. The combination of pattern recognition and feelings of causation and knowing are elevated from perception and sensations to a priori evidence of an underlying verity.

If the lessons of phrenology were to be a guide, you'd think we'd have learned to be more cognizant of the striking variability in the normal size and shape of our brains and the inherent potential for drawing erroneous correlations and conclusions from such observations. And yet, updated measurements of regional variations in brain volume, cell populations, and the thickness and overall density of neuronal connections continue as fundamental tools for correlating anatomy with personality, intelligence, even mental illness. As there is always a new technology on the horizon, criticisms of former methodologies are commonly countered with "That was then; this is now. We have better techniques. Those guys in the past didn't know what they were doing."

To get a feeling for the limits of the measurement approach to mental function, let me present a few peer-reviewed studies using size and quantity of brain cells and connections as the yardstick for various mental traits. Again, my goal is not to criticize a particular study or group of researchers, but to provide an overall sense of the limits of the anatomical approach to understanding the mind. And please note that it is through well-designed studies that reveal new aspects of anatomy that we are given insights into essential limitations of those same methodologies.

Thicker Is Better

A 2011 London *Daily Mail* headline: "Intelligent People Have Thicker 'Insulation' on the Brain's Wires."[8]

As we tend to think of the brain in computer terminology, it has become increasingly fashionable to see intelligence in terms of speed of information processing. We also know that the thicker the insulation around peripheral nerves (the myelin sheath), the faster the speed of conduction of electrical impulses. Combine these two ideas and you have the basis of a 2011 UCLA study attempting to show that the degree of myelin sheath thickness (representing faster processing) correlated with overall intelligence. To test this hypothesis, a group of UCLA neuroscientists led by Paul Thompson decided to study speed of central processing by using a new ultrahigh-resolution diffusion fMRI technique to measure white matter speed of conduction. They also reasoned that a good way to determine the genetic component of this possible correlation between myelin sheath thickness and intelligence was to compare fraternal and identical twins. Identical twins have closely matching IQs, while fraternal twins share only half their genes and exhibit far less similarity of IQ. Demonstrating that identical twins share both a greater similarity of myelin sheath thickness and IQ than fraternal twins would establish a genetic component to intelligence.

As they theorized, the study did reveal that the thicker myelin corresponded with better overall performance on some aspects of IQ testing and that this correlation was seen in the identical twins but not the fraternal twins. The variability in the degree of correlation was seen as evidence of different degrees of genetic contribution for different cognitive functions. According to Thompson, 85 percent of the variation of myelin thickness in the area of the brain that deals with logic, math, and visual spatial skills was genetically determined.[9] However, this correlation between myelin thickness and IQ wasn't uniform. For example, there was no significant correlation between myelin thickness and verbal IQ. In short, some areas showed the expected result while others didn't.

The authors' explanation of this discrepancy: "It may be that performance IQ rather than verbal IQ is more tightly associated with physiologic parameters such as nerve conduction velocity and sensitivity to the level of axonal myelination."[10] Does this argument make sense? For me, it is contrary to basic biological principles—equivalent to saying that nerves in the left arm can be more accurately assessed with nerve conduction studies than nerves in the right arm. This argument could be the ultimate salvage explanation for any inconsistent result. Or perhaps it's a simple problem of the technique not being up to the task. According to a 2011 review article on fMRI diffusion, the technique is still experimental, difficult to use in clinical applications. The review group recommends viewing results based on this technique with a healthy skepticism.[11]

Nevertheless, the authors concluded that "major white matter fiber pathways are highly genetically controlled . . . and linked with intellectual performance." The accompanying press release from the Brain Research Institute at UCLA says that "myelin integrity is an especially promising target for manipulation, because, unlike the volume of grey matter, it changes throughout life. Identifying the genes that promote high-integrity myelin could lead to ways to enhance the genes' activity or artificially add the proteins they code for. Intelligence enhancement in people who just want help passing an exam is, according to Thompson, 'within the realm of possibility.' "[12] Richard Haier, research psychologist at the University of California, Irvine, who has worked with Thompson, said of the study: "Just because intelligence is strongly genetic, that doesn't mean it cannot be improved. It's just the opposite. If it's genetic, it's biochemical, and we have all kinds of ways of influencing biochemistry."[13]

Once upon a time it was thought that the brain was hard-wired. Of course, this "written in stone" perspective couldn't explain how we learn anything. Now we talk of neuro-plasticity—the ability of the brain to change itself. Neural systems are dynamic; learning is reflected in selective changes in the volume of the brain, and corresponding changes in nerve fiber and synapses. As opposed to the UCLA press release arguing that gray matter remains constant throughout life, we have ample evidence that gray matter characteristically increases during new learning. In the monkey-rake experiment, increased brain volume in the appropriate brain region was noted within a week of training the monkeys to use the rake. Memorizing the street layout of London in order to become a licensed cab driver produces a dramatic increase in gray matter volume in the posterior hippocampus, a region of the brain known to be critical for spatial navigation.[14]

In sharp contradiction to the UCLA press release, Thompson's study does acknowledge that our brains are constantly being rewired according to experience. "Myelination varies dynamically throughout life, responding to sensory stimulation or deprivation, nutritional factors and rearing." Curiously, Thompson also presents a counterargument to his own conclusion. "A genetic effect on brain architecture does not imply that environmental factors might not also be responsible for the myelin changes. In many cases, beneficial genes and environmental factors are highly correlated; for example, talented individuals may tend to seek out activities and environments that in turn promote and improve brain function."[15]

Imagine being gifted with a genetic predisposition for enjoying music. You stumble across a gorgeous tuba in a used-instrument store, and spend much of your free time practicing and listening to tuba music. Soon your tuba circuitry in your brain will be enhanced. Connections will function more smoothly, process information more quickly—after all, that's how you get your tuba playing up to speed. In this scenario, a genetically determined predilection toward enjoying music will have led to myelin sheath thickening in your tuba center. This localized brain change will not reflect a genetic component specific to this region or even necessarily to this implied behavior. Before suggesting that enhanced myelin sheaths might improve intelligence, we need to know whether increased myelin sheath thickness was the cause of increased intelligence or, conversely, was an epiphenomenon—the result of a factor that both increased intelligence and incidentally affected myelin sheaths.

Earlier I suggested reading each study as though it were an expert witness at a trial. The Thompson study definitely qualifies as an expert witness, having been peer-reviewed and published in one of the leading neuroscience journals. Thompson's study is now repeatedly cited as an example of how functional brain imaging can provide a good surrogate marker of intelligence *and* as showing that "intelligence is something we inherit."[16] Of course there is a genetic component to intelligence. But that is a far cry from the implicit all-or-nothing genetic contribution implied in such statements. As history has repeatedly warned us, reductionistic statements about the genetics of human behavior carry an enormous potential for misuse and abuse. Keep in mind that similar leaps of logic and misappropriation of isolated bits of questionable scientific data provided the rationale for the practice of eugenics.

Catch-22

Another side of the size-determines-cognition equation is a recent study suggesting that distractibility is caused by "too much brain."[17] University College London researchers used the fMRI to compare the brains of "easy-to-distract" and "difficult-to-distract" subjects. Their measure of distractibility: subjects ranked their own failure to notice road signs or how often they forgot why they had gone to the supermarket. Those with the greatest degree of distractibility were found to have a greater volume of gray matter in the left superior parietal lobe (SPL). Why an increased amount of neurons in a particular region would be associated with impaired attention isn't immediately obvious, but the head researcher, Ryota Kanai, offers an intriguing argument.

As we develop from infants into adults, there is an

approximately 50 percent reduction in the number of neurons in our cerebral cortex. Though the exact mechanisms and reasons behind this "pruning" are not well worked out, the prevailing theory is that pruning gets rid of neural pathways that might be useful early in our development, but are no longer needed when we have established more complex and refined cognitive processing. As we mature, the seldom used or unused neurons are peeled away in order to create a more physiologically and metabolically efficient brain—a metaphoric equivalent might be taking down the scaffolding after a building is completed. Kanai suggests that a greater volume of gray matter may be a sign of a less mature brain, perhaps reflecting a mild developmental malfunction rather than being a sign of increased functionality. According to Kanai, this is consistent with the finding of a greater volume of gray matter in children in comparison with adults and the general observation that children are more easily distracted than adults.

No matter how you feel about the study's findings, you must admire the pure ingeniousness of scientists being able to use a single measurement—brain volume—to document the acquisition of new information, such as a motor skill (as seen in the increased premotor cortex of a monkey learning to use a rake) *or* as evidence of a developmental defect that impairs function. The research team has pulled off the neurophysiological equivalent of having your cake and eating it too. What is particularly clever about this argument is that it can be neither proven nor disproved. As pruning can only be indirectly inferred by painstakingly counting the number of neurons per unit of brain tissue, it cannot be measured in living subjects. Pruning is a statistical concept not detectable at the level of the living individual.

Kanai's finding raises a larger issue—the inherent limitations of using a single data point to correlate brain volume and a particular neurological function. In a University of Texas study, rats were trained to discriminate between similar low-frequency tones. During the learning process, the auditory region for processing low-frequency tones dramatically increased in size, consistent with the idea that learning generates new neurons and/or connections. However, after approximately a month, the expanded areas had receded to their original size even though the ability to discriminate between these tones was retained. In taking a temporal overview of learning, it appears that the acquisition of a new skill may be associated with temporary increase in brain volume, but once the skill is learned, brain volume returns to normal.

The lead author of the Texas study, Michael Kilgard, has an explanation consistent with Kanai's findings. We learn through trial and error. So does our brain; it creates a number of connections in an attempt to solve a problem. Once the optimal solution is achieved, the other, less useful connections are no longer necessary and are winnowed away. With this argument, we can expect pruning to continue throughout life. It is nature's way of getting rid of the detritus of trial and error. The larger implication: brain size both generally and regionally is dynamic, not static. Relying on data from a single point in time is the equivalent of taking a snapshot midway through a close horse race and concluding who the winner will be.

Back to Einstein. It could be argued that his superior intelligence squeezed into a normal-sized brain is evidence for more efficient pruning and superior processing resulting from thought proceeding along wide freeways rather than winding back roads. Perhaps Einstein's brain was huge while contemplating relativity, but that it shrank dramatically once he figured out his space-saving formula, $E = mc^2$. In addition, we have no idea how pruning takes place. Perhaps this is a function of glial cells, in which case the alleged increased number of glial cells in his math center is evidence for ongoing pruning rather than any specific effect of the glial cells. Also, if pruning affects connections, it might have an effect on myelin sheath thickness and integrity. In short, when there are multiple possible explanations of an anatomic finding, we should exercise extreme caution. Any reasonable correlation between regional or global brain size and a specific attribute such as intelligence requires an in-depth understanding of the fluctuations in anatomy and physiology that occur over the lifetime of an individual—a technological tour de force of extraordinarily low probability.

However, as we have good techniques for visualizing presumed brain-volume changes, it remains a fundamental tool of neuroscience. In the wrong hands, the results are often nothing short of spectacular. One particularly glaring example should suffice.

From the May 2011 *Scientific American*: "Religious Experiences Shrink Part of the Brain."[18]

A Duke University fMRI study of several hundred middle-aged men and women was carried out to investigate the role of stress on the size of the hippocampus—a structure central to emotional processing and memory formation.[19] In the past, studies have generally shown that hippocampal atrophy (shrinkage) can be associated with major stress, such as in victims of torture or prisoners in concentration camps. In this study, in addition to assessing overall life stresses, the subjects were also questioned in some detail about their religious beliefs, affiliations, and whether or not they were "born-again" Christians or had had life-changing religious experiences. The study results: no significant correlation was noted between the subjects' self-assessment of their stress levels and hippocampal size. There were, though, some individual differences noted depending upon a subject's religious beliefs. Significant atrophy was seen in individuals reporting a life-changing religious experience, with greater hippocampal atrophy among "born-again" Protestants, Catholics, and those with no religious affiliation, compared with Protestants not identifying as "born-again."[20]

The authors' conclusion: hippocampal atrophy in selected religious groups might be related to stress! They theorize that some individuals in the religious minority, or those who struggle with their beliefs, experience higher levels of stress. This causes a release of stress hormones that are known to depress the volume of the hippocampus over time. This might also explain the fact that both nonreligious and some religious individuals have smaller hippocampal volumes.

If you were reviewing this study for a prestigious popular science journal such as *Scientific American,* what would you write? Here's what the director of research of the Center of Integrative Medicine at Thomas Jefferson University in Philadelphia, Andrew Newberg, M.D., wrote:

> The authors certainly pose a plausible hypothesis. The authors also cite some of the limitations of their findings, such as the small sample size. More importantly, the causal relationship between brain findings and religion is difficult to clearly establish. Is it possible, for example, that those people with smaller hippocampal volumes are more likely to have specific religious attributes, drawing the causal arrow in the other direction? Further, it might be that the factors leading up to the life-changing events are important and not just the experience itself. Since brain atrophy reflects everything that happens to a person up to that point, one cannot definitively conclude that the most intense experience was in fact the thing that resulted in brain atrophy. So there are many potential factors that could lead to the reported results. (It is also somewhat problematic that stress itself did not correlate with hippocampal volumes since this was

one of the potential hypotheses proposed by the authors and thus, appears to undercut the conclusions.) One might ask whether it is possible that people who are more religious suffer greater inherent stress, but that their religion actually helps to protect them somewhat. Religion is frequently cited as an important coping mechanism for dealing with stress.

His final conclusion:

This new study is intriguing and important. It makes us think more about the complexity of the relationship between religion and the brain. This field of scholarship, referred to as neurotheology, can greatly advance our understanding of religion, spirituality, and the brain. Continued studies of both the acute and chronic effects of religion on the brain will be highly valuable. For now, we can be certain that religion affects the brain—we just are not certain how.[21]

Forget the possibility of underlying agenda—Newberg is the author of *How God Changes Your Brain: Breakthrough Findings from a Leading Neuroscientist*.[22] Disregard the shaky logic, such as showing that overall levels of stress didn't correlate with hippocampal atrophy, yet suggesting that stress was likely to be the cause of the hippocampal atrophy seen in some subjects who reported life-changing religious experiences. What is most startling is Newberg's alternative hypothesis that the size of the hippocampus might be a reflection of an underlying tendency toward specific attributes. This is like arguing that brain size might determine whether we are Protestant, born-again, or atheist. If this fine-grained attribution of religious choice to variations in anatomy can be considered science, phrenology is gospel.

In 2007, the *Royal Society* of England, in its flagship biological research journal, issued an extensive critique of twenty-five years of published results on brain size and behavior. "We all know that correlation does not demonstrate causation but causation is the context in which the results are invariably interpreted."[23] The society pointed out that neuroscientists disregard the lessons of history, remain ignorant of prior and present studies asking the same questions, generally persist with inadequate data collection, fail to carry out suitable confirming studies despite their availability, limit their correlations to those that confirm their hypotheses, and cite correlation as evidence of causation.

Have neuroscientists heeded the Royal Society's criticisms? I'll let you answer that by offering one last example of an anatomical technique being offered as a potential major neuroscience breakthrough.

Wired to the Max

In December 2010, the *New York Times* reported that a group of Harvard and MIT researchers had developed a method to uncover the brain's entire wiring diagram. To perfect the technique, they are starting by slicing a mouse brain into ultrathin slices that can be seen under electron microscopy. By taking detailed photographs of each slice and then reassembling the composite images, every connection between every nerve cell within the nervous system will be revealed. Their eventual goal is to transfer this technique to humans in order to "build a full map of the mind."[24] The researchers say this project, analogous to the Human Genome Project, will reveal the mental makeup of a person, including how memories, personality traits, and skills are stored. In September 2010, the National Institutes of Health handed out $40 million in grants for them to pursue the "Human Connectome Project."

To get a feel for the enormity of this project, consider the numbers involved. To date, the only available anatomic wiring diagram is for a microscopic worm (*C. elegans*). Mapping three hundred neurons and their seven thousand connections was a Nobel Prize–winning effort that took more than a decade to complete. A mouse brain has 100 million neurons, each with a large number of connections. It is estimated that the amount of information contained in a one cubic millimeter of mouse brain would require 1 petabyte (1,000 terabytes) of memory— the same amount of data storage necessary for Facebook to store 40 billion photos. The human brain with its 100 billion neurons and 100 trillion synapses would require a 1-million-petabyte storage facility to contain the images. According to Jeff Lichtman, Harvard professor of molecular and cellular biology and collaborator in the connectome project, "The world is not yet ready for the million-petabyte data set the human brain would be, but it will be."

At first glance this sounds like an enormous challenge for data collection and storage. But even if these obstacles can be overcome, what might we learn? The researchers assume that a snapshot of the brain at any moment in time would give you a long-term picture of the brain's wiring. Yet they readily acknowledge that the brain is dynamic and ever-changing in its connections. However, the prospect of a wiring diagram for the brain is just too enticing.

Gary S. Lynch, neuroscientist at the University of California, Irvine: "Lacking a blueprint, we're never going to get anywhere on the most profound and fun questions that drew everyone to neuroscience in the first place: what is thought, consciousness?"[25]

Okay, let's assume that we arrive at the theoretical day when we have a complete wiring diagram of every connection for every neuron at every instant. As the technique requires postmortem deconstruction and reconstruction of ultrathin brain slices, interviewing the subject isn't possible. We would need to find some anatomic equivalent of the expression of a thought or the presentation of a mood. But nasty thoughts don't look like angry neurons; a good mood isn't represented as a smiley in a synaptic vesicle. Thoughts aren't accompanied by labels or hovering cartoon-style balloons. We are still left with the same problem—we can know the contents of consciousness only by communicating directly with the subject. And the Human Connectome Project process does not allow us to communicate directly with the subject.

If Lynch is right, and many scientists are drawn to neuroscience in order to pursue an understanding of thought and consciousness, university career counselors should take note that a career unraveling brain wiring will not bring our future scientists any closer to answering these questions. The techniques of the Human Connectome Project, if successful, might provide a spectacular and valuable blueprint for investigating how our brain parts interact. But believing that knowledge of brain wiring can tell us the nature of our consciousness is like predicting what sound will come out of a set of speakers by looking at a wiring diagram of the component parts. Even if you had perfect knowledge of how bits of information are converted into sound waves, you wouldn't buy a set of stereo speakers based upon a wiring diagram. The wiring does not predict the quality of conscious experience.

And yet the Human Connectome Project has been greeted with evangelical fervor. Consider the Technology Entertainment, Design (TED) presentation of Sebastian Seung, MIT professor of computational neuroscience and co-creator of the connectome.

To begin his talk, Seung asks the audience members to chant along with him, "I am my connectome." Seung argues that thoughts, personality traits, and memories are stored in connections between neurons. To establish the truth of this claim, he proposes that we should be able to directly read out a memory from our brain connections. "Memories are stored as a series of synaptic connections inside your brain," he says. "One way to test this theory would be to look for a sequence of synaptic connections within a connectome. The sequence of the neurons that we recover [with the connectome] will be a prediction of the neural activity that is replayed in the brain during memory recall. And if that were successful, that would be the first example of reading a memory from a connectome."

Missing from Seung's enthusiastic forecast is the methodology for knowing the contents of a thought by looking at its trail of synapses. But never mind. Seung goes on to say that if cryonic freezing of the brain preserves the connectome, memories could be resurrected. He then finishes on a high note: "Connectomes will mark a turning point in human history. . . . Eventually these new technologies will become so powerful that we will use them to know ourselves. I believe the voyage of self-discovery is for all of us."[26]

Let's recap. Seung is suggesting that we can directly read out memories from our wiring, that these memories might be preserved after death if our neural circuitry could be spared from postmortem changes, and that his research will be a turning point in human history. I cannot imagine a better example of faith-based magical thinking. Seung's faith: The mind and its contents must be fully represented in our synapses and their connections.

The Human Connectome Project may well yield important information for understanding diseases and mental disorders. But it is a huge mistake to believe that unveiling anatomy is equivalent to revealing thoughts and memories. Understanding anatomy is necessary but not sufficient for understanding the mind.

12 • Moral Character—Fact or Fiction?

The first men probably did not know where their
thoughts ended and the consciousness of beasts began.
—Doris Lessing, on D. H. Lawrence's "The Fox"

W hich of the following statements seems most reasonable?

Champadi Raman Mukundan, the founder of the EEG
technique that helped convict the woman from Mumbai of
murder: "Man is not destined to be controlled by nature. Man
is destined to control nature."

Stephen Hawking: "Philosophy is dead."[1]

Freeman Dyson, theoretical physicist: "Science is not a collection of truths. It is a continuing exploration of mysteries."[2]

Early in my training I heard neurophysiologist John Eccles
describe his Nobel Prize–winning studies on the synapse.
After presenting his data to the scientists crowded into the
small University of California, San Francisco, conference
room, he turned off the slide projector, stepped out from behind the podium, sat on a nearby desktop, and quietly told us
that the mind and the brain were two separate entities. What I
most vividly recall is the sense that turning off the projector
and walking from the podium to the desktop intentionally
demarcated the conclusion of the scientific portion of his talk
and implied that his remarks about the mind were speculative
rather than based on hard science. The audience understood

the difference. After all, Eccles was a neurophysiologist, not a philosopher; no one seriously entertained the notion that the two fields had much if anything in common, or that his metaphysical utterances were anything more than the idle musings of someone stepping outside his field of expertise. (About the same time, I heard a prominent British philosopher quip that the brain was a knot at the end of the spinal cord to keep it from unraveling.)

John Eccles should serve as a cautionary lesson to anyone who uses brain science to explain the mind. Eccles's seminal contributions to basic neurophysiology have paved the way for our understanding of synaptic transmission. Meanwhile, his speculations on mind-body dualism have been relegated to the dustbin of outdated theories. Fortunately, Eccles had the common sense not to present his personal beliefs as hard science.

Times change. Combine the rapid development of highly sophisticated tools for investigating the brain with the assumption widely held in the neuroscience community that the brain is the mind (or creates the mind), and you have the ideal environment for perpetuating the growing belief— among both scientists and the lay public—that neuroscience will be able to provide answers to age-old philosophical conundrums. Over the last few decades, neuroscience, a once low-key laboratory subspecialty with little practical application or public awareness, has been transformed into a high-visibility field with an authoritative stronghold on the intellectual community.[3] The good news is that the surging interest in the brain is attracting some of our smartest students, is prompting better funding, and, most important, is yielding profound insights into brain function. The unfortunate downside of neuroscientists assuming the role of philosopher-kings is the natural tendency to make speculations and generalizations that are beyond their training and expertise.

Overlooked or ignored in their enthusiasm for scientific answers to philosophical issues is the hard-to-swallow reality that the failure to resolve many philosophical arguments about the mind isn't just due to a lack of scientific ingenuity. Many of the issues are themselves riddled with false assumptions, irreconcilable contradictions, frank paradoxes, and metaphysical issues that cannot be scientifically addressed.

As a prime example, briefly consider the question of free will. Though our mind quickly rushes forward to think of all the various philosophical arguments, take a moment to appreciate the origin of the concept. We all experience a sense of self, agency, effort, choice, and causation—involuntary mental sensations that collectively create a sense of both having and making choices. It is unimaginable that without these feelings, we would ever dream up the idea of free will any more than a tree contemplates the meaning of heartbreak. All philosophical positions about free will emanate from an inbuilt desire to explain an involuntary sensation. No matter how profound our thinking might be, we are saddled with the built-in paradox of automatic and "hardwired" mental states telling us that we are free to choose and act on every whim while science tells us that all actions have antecedent physical causes.

Spending more than a few minutes on the well-reasoned but contradictory arguments for the presence or absence of free will produces intellectual vertigo. No argument stands out as being above paradox and logical inconsistency.[4] Even the phrase "free will" is redundant; what other kind of will would there be? But what should we expect from a question triggered by involuntary sensations and shaped by our individual perceptions?

In this chapter, we'll take a look at some major areas of neuroscience in which the uncritical commingling of conflicting philosophical and scientific principles has led to some extraordinary claims about subjects as diverse as moral judgments, the nature of character and wisdom, and what is "real."

The Anatomy of Morality

The cognitive science and moral philosophy communities tell us that morality and character have their roots in biology. This is hardly surprising. Where else would such qualities be found? Unless you believe that principles of morality exist independently in some ideal Platonic sphere, you would expect that morality and character arise out of our passions, beliefs, desires, thoughts, and experience—all of which are reflected in our biology.

If we conclude that our morality is driven solely by our innate biology, we are faced with a fairly dim view of the human condition. If, on the other hand, we deny the major role that biology plays in the determination of our morality and character, we are swimming upstream against compelling contrary data. Of course, in practice, most of us believe in neither extreme—morality and character are the complex interaction of nature and nurture. The problem with this commonsense view is that it doesn't allow for the neat categories of classification of behavior necessary for good scientific research.

Consider the subject of fairness. Our ability to determine fairness and make moral choices has long been considered a key feature separating man from the rest of the animal kingdom. The underlying assumption is that moral choices arise out of conscious decision making singular to mankind. But evidence suggests that a sense of fairness extends a considerable distance down the evolutionary tree, and its manifestations are easily observed in ravens, wolves, coyotes, domestic dogs, capuchin monkeys, and chimpanzees—animals historically excluded from consideration for having a "moral life" or reason-based character.

Chimpanzees and bonobos will voluntarily open a door to offer a companion access to food, even if they lose part of it in the process. "Innocent bystander" ravens have been observed attacking a raven that has stolen food from another, even though there is no gain for the attacking ravens. Capuchin monkeys can play an animal version of the ultimatum game, using tokens to get food for others, even though it means less food for the individual.

At the risk of overinterpreting nonverbal animal behavior, the underlying common denominator for acts of "fairness" appears to be the social nature of the species. It is as though social animals have evolved a sense of equity and inequity in order to optimize their survival. However, we are still left without a clear sense of what "fairness" is at the level of animal experience. A cynic might see acts of presumed fairness as nothing more than the animal kingdom version of a con man setting up his mark for a future score. An animal lover might see the behavior as evidence for compassion between brethren. One point of agreement remains: what looks like moral behavior does not require formal language or complex processes of reasoning.

Extrapolating to humans, many experts in the field now feel that our moral judgments are also primarily triggered by underlying emotions and mental feelings, with our conscious minds providing after-the-fact rationalizations for our behavior. According to University of Virginia psychologist Jonathan Haidt, "The emotions are, in fact, in charge of the temple of morality, and . . . moral reasoning is really just a servant masquerading as a high priest."[5] Haidt argues that after-the-fact reasoning isn't to seek out the truth of a position, but to convince other people (and also ourselves) that we're right. Some researchers take the further step and conclude that moral judgments are analogous to aesthetic judgments. In the same way that we know that a cup of coffee tastes good or that a painting is beautiful, we experience a visceral knowledge of whether or not a decision is morally correct.

The most famous and commonly discussed thought experiment for determining the role of emotions on moral decision making is the classical trolley car experiment. The abbreviated version: A trolley car is rushing down a railroad track at high speed. On the track are five people who are in imminent danger of being killed by the oncoming train. However, if you pull a lever, the train will be diverted onto a side track where one person is standing. When asked, the vast majority of people indicate that they would pull the lever, knowingly killing one to save five lives. However, when the conditions are changed and you are asked to push a person onto the track to stop the train before it strikes five people, few are willing to physically shove the person onto the track.

The multiple possible interpretations of this experiment have spawned an academic industry whimsically referred to as "trolleyology." The prevailing interpretation is that we can readily make a rational decision when we aren't physically involved, but we have difficulty overcoming more basic emotions up close and personal. Even though the outcome is understood to be the same, innate disgust and revulsion take precedence over a utilitarian "kill one to save five" reasoning.

Confirmation of the role of biology on moral decisions comes from studies on psychopaths—criminals with recurrent antisocial behavior unassociated with any feelings of remorse. Those with demonstrable damage to brain regions crucial for properly processing emotions tend to see pushing the man onto the tracks as morally equivalent to pulling the switch. From the utilitarian perspective, freed from the emotional pulls of empathy, revulsion, and disgust, the psychopath is more likely to consistently apply the same line of reasoning—kill one to save five—than most of us. The decision is mere calculation, without the intrusion of contrary feelings.

Others have used the same research material to arrive at a contrary viewpoint—that emotional experiences such as disgust and revulsion may follow from moral decisions as opposed to preceding and guiding them. This about-face opinion claims that psychopaths make the same kinds of moral distinctions as do healthy individuals. Normal social emotional processing does not appear necessary for making these kinds of moral judgments. The researchers' bottom line conclusion: "Psychopaths know what is right or wrong, but simply don't care."[6]

The social and legal ramifications of how we interpret such studies are enormous. If we view a psychopath as biologically incapable of controlling his violent behavior, we will treat him quite differently than if we see him as a hardened criminal who simply doesn't care. But are we capable, on the basis of cognitive science studies, of making such a decision? Can we make such distinctions without first addressing the problem of the nature of intention? Are such laboratory experiments indicative of how we would react in everyday life? Can we identify the moral component of a decision or action with brain-imaging studies?

Let me present a couple of personal examples. As a liberal arts major in college, I had only a single introductory course in biology. Having skipped over frog dissection day, I was utterly unprepared for the first day of med-school anatomy class. Rubber-sheet-covered bodies filled the dissection room; beyond, through several large windows, was a magnificent view of the Golden Gate Bridge. Following the head instructor's introductory paean to the joys of gross anatomy, we were instructed to pull back the sheets. All at once, twenty-five dead bodies came into view. The other students grabbed their scalpels and went to work. I stood motionless, flooded with a sense of revulsion, strangeness, horror, and anxiety. I can remember wanting to leave the room and medical school. Instead, I accused my new lab partners of lack of respect for our cadaver by leaving her face exposed. To make my point, I leaned over and covered her face with a towel. A month later, my moral indignation subdued by the familiarity of routine and my growing fascination with the human body, I found myself propping up my elbow on the dead woman's chin while reading from the dissection manual.

Evolutionary biologists suggest that disgust is a primary emotion necessary for survival. For instance, a bad smell or revolting sight will steer us away from eating spoiled meat. Yet if we were studying how my squeamishness, disgust, and revulsion at the sight of the cadavers generated moral indignation, a first challenge would be to determine that such feelings were primary as opposed to being triggered by yet other mental states such as a sense of unfamiliarity, strangeness, existential dread, or fear of death. Such a complex interaction of mental states is the very definition of life experience, but it would show up on functional imaging only if these contributing mental states were all activated *at that particular moment*.

Once again we must return to the problem of baseline measurements.

Now Is Not Forever

Think of all the mental and motor skills involved in learning to play the piano—from the position of your elbows to the angle of attack of your fingers. Once you become an accomplished pianist, you no longer need to be aware of the individual elements collectively necessary to play a particular piece. Being relieved of the need to constantly remind yourself how to hold your hands and where to place your feet is translated physiologically into less overall mental effort than when you were first learning to play. Less effort equals less metabolic demand, which translates into a lessened likelihood that these areas will light up on functional imaging scans. (This is the same general principle responsible for the transient increases in gray matter volume while acquiring a skill. Once a skill is learned, unneeded neural connections may be winnowed away.)

If you have learned to sit upright with your arms extended at a certain length, and no longer consciously think about your posture, this subliminal knowledge is more likely to result in low-level brain activity present whenever you are at the piano rather than as a transient burst of increased activity that can be detected on an fMRI. This is the same problem of detecting baseline brain activity that we saw with the Pete and Mike example in chapter 9.

If we think of a moral decision as being partially determined by immediate circumstance, and partially by ongoing biological predispositions and the cumulative effect of life experiences, the entire notion of the fMRI providing a complete picture of the contributions to a moral decision becomes suspect. If I had been wearing a Superscanner on the first day of anatomy class, the fMRI would have singled out those regions of the brain that were activated at the moment of my personal revulsion and sense of moral indignation, but would not have accurately detected my antecedent existential concerns or fears of death that hover in the background as chronic low-grade aspects of my personality.

Fair Is Fair

A further problem in studying the biology of morality is in how we conceptualize a particular behavior. I was recently rear-ended on a freeway off-ramp by a Ford F-150 truck. No one was hurt, but the trunk of my relatively new midsize car was caved in. Normally, when I have been wronged out of stupidity or carelessness, I've been irate and morally indignant. However, the driver of the truck was a young mother of two young children, both of whom were in the front seat, crying, wailing, and screaming. The mother explained that she had been breast-feeding one child (while driving) and her cell phone went off! She couldn't find her driver's license, and her insurance coverage had expired. To my utter surprise, I patted her on the shoulder and said, "Don't worry, you'll be okay." I was as stunned by my gesture and comment as she was. To this day I remain puzzled by my behavior, and, central to this discussion, cannot figure out any scientific way to dissect it out into something measurable and analyzable.

Perhaps I had a minor concussion and was confused. Or I was relieved that I wasn't hurt. What I recall popping into my mind just after consoling her was the realization that life is intrinsically unfair and that this woman's life was probably harder than mine. While waiting for the tow truck, I wondered about the nature of fairness. I had previously considered the concept of fairness as a purely cognitive decision—the balancing of individual rights and responsibilities. Now I wondered if fairness might be an expression of a more deeply rooted sense of one's place in the world. Possible underlying triggers for my spontaneous reaching out could have included my understanding of luck and misfortune, my personal sense of gratitude and entitlement, and regard for the lot of others.

This issue was at the heart of a recent debate over how best to choose a Supreme Court justice. President Obama said, "I view that quality of empathy, of understanding and identifying with people's hopes and struggles, as an essential ingredient for arriving at just decisions and outcomes."[7] On Bill Bennett's *Morning in America* radio show, former Republican National Committee chairman Michael Steele said, "I don't need some justice up there feeling bad for my opponent because of their life circumstances or their condition and short-changing me and my opportunity to get fair treatment under the law." Richard Epstein, a legal scholar and professor of law at the University of Chicago, seemed to agree. "Empathy matters in running business, charities and churches," he said. "But judges perform different functions. They interpret laws and resolve disputes. Rather than targeting his favorite groups, Obama should follow the most time-honored image of justice: the blind goddess, Iustitia, carrying the scales of justice."[8]

Where you position yourself between these opposing viewpoints depends upon the degree to which you see fairness as a mental sensation versus a conscious determination. For me, a major component of fairness is the ability to put oneself in another person's shoes both intellectually and emotionally. For those who believe that fairness is arrived at through conscious deliberation, a relative lack of empathy is interpreted as a positive trait—the ability to put aside personal feelings and be objective. Though this is pure speculation, I suspect that this fundamental difference in how to think about fairness is a critical component of the global increase in vicious partisan politics. One man's fairness is another's injustice, be it abortion, immigration, the death penalty, or taxes.

Unfortunately, there is no scientific methodology that can show us the true nature of fairness. There is no center of the brain that can be stimulated to make one feel a sense of justice, no area of the brain that will light up when making a moral choice that can be traced back to "fairness" neurons. In the end, we are left to our own involuntary feelings and perceptions to decide on the nature of an abstract concept that is likely to determine the future of our civilization.

But that's not how UCLA neuroscientist Sam Harris sees it. Harris, a leading voice in the so-called new atheist movement, believes morality and fairness can be pinpointed in the brain. In his introduction to *The Moral Landscape,* Harris claims:

> Questions about values—about meaning, morality, and life's larger purpose—are really questions about the well-being of conscious creatures. . . . Values translate into facts that can be scientifically understood. . . . There are facts to be understood about how thoughts and intentions arise in the human brain; there are facts to be learned about how these mental states translate into behavior; there are further facts to be known about how these behaviors influence the world and the experience of other conscious beings. . . . Facts of this sort exhaust what we can reasonably mean by terms like "good" and "evil."[9]

Harris is confident that science will allow us to "identify aspects of our minds that cause us to deviate from norms of factual and moral reasoning."[10] He then makes the leap to believing that such knowledge is the path to knowing which actions will best serve our collective well-being. But there is no fact about the mental states of the young mother and myself that can be converted into a scientifically optimal mode of behavior.

Even if Harris has perfect neuroscience to back up his claims (which he doesn't), we are still left with how to use science to tell us what the good life is. Imagine a hypothetical society in which pure rationality both existed and prevailed (a low-likelihood scenario) and 100 percent of subjects would push the man onto the railroad track to save five lives. On the surface, this would seem entirely consistent with pulling the lever to divert the train and would be the mark of reason overcoming baser irrational instincts. One could even see this society as maximizing our collective well-being.

But would you want to live in a society where everyone disregarded personal feelings for the individual, no matter how irrational, in order to live according to the algorithms of utilitarianism? How would you look upon your friends and neighbors if you knew that their decision to help you in an emergency would be predicated solely upon the collective good as determined by laboratory experiments? There is no moral right or wrong to these questions (nor am I advocating one kind of society over another). Your answers to these questions are reflections of personal taste and your preference for the type of society in which you would choose to live.

To take Harris's argument to its logical extreme, imagine a time when science revealed that a love affair was a matter of sky-high oxytocin levels, your intelligence was determined by the thickness of your myelin sheaths, and your sense of purpose was nothing more than limbic-system activation. Some might find solace and moral direction from this view of human nature, while others might be horrified that they were expected to optimize their lives by taking their moral lessons from such facts.

If a goal of science is to reveal what will provide us with the maximum well-being, we still need to understand what the term "well-being" means and whether "us" refers to each individual (libertarianism) or society as a whole (utilitarianism). Even if we have overwhelming evidence that cigarette smoking causes lung cancer, gambling can be a destructive addiction, TV rots your brain, and the Internet plays havoc with your attention span, there remain huge philosophical issues as to how people should/ought/might/must lead their lives.[11] To believe that neuroscience can provide these answers is to believe that messy human behavior can be reduced to scientific facts. I cannot imagine a more faith-based, scientifically untestable point of view.

Character Study

Fool me once, shame on you. Fool me twice, shame on me.

Equally shortsighted is the role neuroscience has assumed in redefining the nature of character. One of the central tenets of being a good person is being of "good character." It is fundamental to everything from child rearing to the Boy Scouts to learning how to be a member of a family, a team, a corporation, or the Donner party. Character is at the top of the list of traits that we use to judge each other, and how we assess ourselves. As Heraclitus once said, "Character is destiny." But what is character? How can the self—a brain-generated virtual construction that is constantly changing—be said to have character?

According to the latest in cognitive science, character is at best a partial truth; our behavior can be dramatically and involuntarily affected by circumstance. One of the more commonly cited examples is that the smell of bakery goods increases one's likelihood of being generous. In a classic study, a stranger was more likely to give change for a dollar if approached outside a "fragrant bakery" than outside a "neutral-smelling dry-goods store."[12] Similarly, the presence of a clean (usually citrus) scent has been shown to increase volunteers' degree of virtuous behavior. In one study, a group of volunteers showed a much greater degree of reciprocal trust and charitableness if they were in a room that had recently been sprayed with citrus-scented Windex as opposed to a room with no scent.[13] Perhaps the most damning rebuttal of intrinsic character determining morality is the study showing that seminary students on the way to give a lecture on morality would not stop to help a stranger in need if they believed they might be late for the lecture. Other notorious examples are the Stanley Milgram study from the sixties showing that given sufficient incentive, volunteers would be willing to apply potentially lethal electric shocks to test subjects, and the Philip Zimbardo prisoner study from Stanford, where students, when assigned the roles of guard and prisoner, assumed the character traits common to real guards and prisoners, including the tendency for "guards" to physically assault "prisoners."

Such evidence of our ability to be dramatically influenced by circumstance has generated a new philosophical concept—"situationism." Gilbert Harman, a Princeton philosopher, recently wrote that "ordinary attributions of character traits to people are often deeply misguided. It may even be the case that there is no such thing as character, no ordinary character traits of the sort people think there are, none of the usual moral virtues and vices."[14] John Doris, Washington University professor of philosophy and neuroscience, recommends abandoning thinking about human behavior and moral capacities in terms of broad traits of character such as "honesty," "bravery," and "self-reliance."[15]

I doubt that anyone seriously believes that character is exclusively a reflection of conscious decisions; we are all aware of circumstances prompting unexpected "out-of-character" behavior. But character is more than the isolated moment. It is the accumulation of all experience and personal biological traits. If we go back to the notion of a hidden layer containing all of our predisposing biology and experience, it is easy to see that circumstance will be an input that can shift the scales and trigger a change of character.

On the other hand, character isn't a specific brain function or biological property; it is a concept we use to describe how all the various aspects of our biology and experience coalesce into some degree of predictability of behavior. Character is the assignment of probabilities—the likelihood that an individual will be hardworking, trustworthy, and loyal or will go postal and shoot his employer. To argue that character doesn't exist is to look at the wrong level of explanation of behavior, as mistaken as arguing that pain doesn't exist because it cannot be located in neurons. If we want to understand individual traits, such as why one person is more honest than another, we can look to genetic studies on identical twins raised apart or to fMRI studies to see what areas of the brain are activated when a subject decides to be honest or dishonest.[16] Via such studies we can get closer to understanding the basic mechanisms at work when one makes an honest or a dishonest statement. However, this is a far cry from looking for a specific trait at the level of brain function. There is no honesty center or circuitry.

Though character is an abstraction—a semantic device we use to judge past behavior and predict the likelihood of future behavior—it still exists in the sense that it directly affects our behavior. Case in point: my understanding of my character and the decision to act in this manner directly bear on how I write the following sentences. In writing this book I am acutely aware of having strong negative feelings toward certain unjustifiable neuroscientific claims. At the same time, one of my underlying themes is to remain open-minded and consider alternative possibilities that conflict with one's own viewpoint. My sense of obligation to my own self-image operates as an input into the hidden layer that outputs my comments. Even if my sense of my own character is utterly wrong—an unreliable personal fiction that I've tacked onto a virtual self—this imagined self-image plays a real role in how I behave, just as the belief in flying saucers might dictate the size of the welcome mat on a believer's front door. Nietzsche once said, "Active, successful natures act, not according to the dictum know thyself, but as if there hovered before them the commandment: will a self and thou shalt become a self."[17]

As mistaken as the belief that character is nonexistent because it isn't a specific brain function is the contrary idea that character can be located within specific neural circuitry. Consider these recent headlines: "Brain Scans May Identify Slackers"[18] and "Optimism Is a Brain Defect, According to Functional MRI Scans."[19] But such simplistic claims pale in comparison with the notion that character traits can be physically altered through direct medical intervention.

Researchers at the Weizmann Institute of Science in Israel conducted an fMRI study of the relationship between courage and fear. Participants were categorized as "fearful" or "fearless" depending on how they responded to a snake-fear questionnaire. They were then asked to move a live snake (nonpoisonous) closer to their bodies. Those "fearful" participants who were able to overcome their innate fear of being close to a snake had greater levels of a specific localized brain activity.[20] The lead researcher's conclusion: it might be possible to therapeutically enhance this brain activity to increase one's level of courage.[21]

Traits (aspects of character) arise out of the interaction of myriad hidden-layer elements, but don't exist at the level of cells and synapses. Character isn't pure physiology although it is prompted by inherent biological tendencies. Neither is it purely dictated by circumstance. Rather, it is a concept, a descriptive vehicle for assigning probabilities of behavior that arises out of the complex interaction of an organism with its environment. To conclude that character traits either don't exist or, conversely are primary brain functions potentially amenable to therapeutic intervention is yet another example of how looking at the wrong level of explanation of behavior results in profoundly misguided views of human nature.

Wising Up to Intelligence

Occupying the top rung in the hierarchy of character traits is wisdom. More than any other value, most would consider wisdom to be the finest quality of a good mind. So how does neuroscience weigh in on this lofty subject? A prominent British neuroscientist recently proposed a new set of tests for intelligence, bundling together a dozen measurements that he believes cover the broadest range of cognitive skills and most extensive testing of different anatomic areas of brain function. He has called these tests "the 12 pillars of wisdom," and has suggested that these measurements "might be called the ultimate intelligence test."[22]

For the moment, put aside the age-old controversies regarding both the definition of intelligence and the problems of standardized testing. Let's assume that these tests give a perfect indicator of overall intelligence (whatever that means). Reducing a complex quality—wisdom—to a set of numerical values fails to take into account other necessary ingredients—humor, sense of irony, feelings of empathy, regard for integrity and fairness, to name but a few traits high on my list of what makes someone wise. But let's put these considerations aside as well and look at two of the tests that are part of the "12 pillars of wisdom" battery. One of the tests evaluates visuospatial skills and another measures the ability to rotate images in your mind.

Imagine two people of identical intelligence on everything

except these two tests of visuospatial orientation and function. Assume one person does significantly worse on these two tests but the same on all the others. Should we conclude that he is less wise then someone with better spatial orientation? In an era when early life aptitude and intelligence testing plays an increasingly important role in determining where and how a child will be educated, do we really want to classify a child as more or less wise depending upon the speed with which she can rotate the image of a cube in her mind? In his recent book *The Mind's Eye,* Oliver Sacks describes his own visuospatial deficits, including an inability to recognize faces. Is this somehow evidence that Dr. Sacks is not a wise man? Even if spatial orientation is perfectly correlated with general intelligence, being able to rotate a cube in one's mind doesn't seem to be critical to determining the best approach to world peace, mitigating global warming, avoiding marital strife, or choosing which high school is best for your child.

For many years I took care of a young man with subnormal intelligence due to birth trauma and a difficult-to-control seizure disorder. He had little schooling, lived in the San Francisco Tenderloin, had occasional odd jobs with Goodwill and the circus, hung around with a rough crowd, and had some minor skirmishes with the law. One day, he appeared in the office and told me he was thinking of getting married to a woman who suffered from brain damage from an auto accident. I still remember his haunted look when he asked if someone "slow" married someone else who was "slow," would their child also be "slow"? I was particularly struck by the way that he kept circling around the word "slow," repeating it with slightly different intonations as if trying to get a handle on what the word meant in a larger context. I explained that brain injuries weren't passed down to your children. For some time he sat quietly, head down, hands on his knees. Then he looked up and asked, "Would my being slow be a problem for the kid, I mean, would it cause him problems? After all, I want to do the right thing by him." To me, that is wisdom.

Equating intelligence with wisdom is not wisdom; it is hubris. It is an attempt to isolate and spotlight one aspect of the mind—intellectual prowess—and make that into the defining feature of a man. I have to say, at the risk of being unnecessarily cynical, that equating intelligence with wisdom is a spectacularly self-serving way of converting one's own presumed intellectual horsepower into the role of moral superiority. It is this thinly disguised self-congratulatory posture that prompts some scientists to presume that they have the inside track in determining moral values, establishing a "theory of everything," or arguing that "philosophy is dead."

Get Real

Perhaps the most extraordinary incursion of neuroscience into the philosophical domain is the belief that it can establish what is "real." An October 25, 2010, BBC News headline stated, "Libido Problems 'Brain Not Mind.'" The caption beneath the accompanying photograph of a concerned young woman read, "Altered brain blood flow may explain a lack of desire, scientists believe." The headline and photograph were referring to a study from Wayne State University that was presented at the annual meeting of the American Society for Reproductive Medicine. The lead researcher, Michael Diamond, M.D., was interested in knowing whether there were detectable differences in levels of brain activity in women with so-called normal sex drive in comparison with those who received a diagnosis of "hypoactive sexual desire disorder" (HSDD). Dr. Diamond showed both groups erotic videos. In the control group, the erotic videos triggered increased activity in the insular cortices—parts of the brain believed to be involved in the processing of emotions. Those with the diagnosis of HSDD showed no increased activation.

Dr. Diamond's conclusion: "Being able to identify physiological changes provides significant evidence that it is a true disorder as opposed to a societal construct. . . . The study provides a physical basis suggesting that it is a true physiological disorder."[23] The overriding belief that neuroscience has the tools to redefine the human condition has prompted a researcher to believe that increased blood flow to an area of the brain is capable of determining "what is real." If a region of the brain could be seen to light up on an fMRI when subjects imagine three-legged Martians surfing on a sea of concrete, would that make the Martians "real"? A concrete sea "real"? And what would constitute a "false" mental disorder?

How extraordinary that millennia of philosophical contemplation of the nature of reality can be swept aside so that the new definition of "real" is solely dependent upon an algorithm-driven computer-constructed brain image. Further, it is well known that metabolic brain changes occur for a variety of reasons. If you are depressed, overworked, underpaid, generally misanthropic, or having a bad hair day, watching a video of a beautiful and exuberant couple doing an erotic tango may not push your emotional buttons. If not, the blood flow to your insular cortex will not increase. This is self-evident; all aspects of one's psyche contribute to our emotional responses. A lack of increased blood flow in emotional regions of the brain tells you absolutely nothing about the underlying cause or causes. To distinguish between psychological and physical on the basis of changes in brain activity is nothing short of a belief in mind-body dualism gone wild.

The most bothersome aspect of this study is the cavalier interpretation that an fMRI finding is evidence of disease without addressing the long-term implications of labeling a behavior (lack of desire) as a "physical condition." The misappropriation of a perfectly good technique for localizing brain activity is, in effect, being used to tell a patient that she is "sick" without any clear idea of the underlying mechanisms, or even what the label means in terms of possible treatments. If a patient believes that there is something "wrong with my brain," the effects can be disastrous. Anyone who has been told of a possible abnormality on a lab test knows how hard it is to shake off that disturbing knowledge even when repeat studies turn out to be normal.

Sadly, such misuse of fMRI findings to establish the "realness" of a controversial disorder is omnipresent. Look at the muscle-pain disorder fibromyalgia. Despite strong convictions on all sides, nobody knows whether fibromyalgia is a primary medical condition, is part of a larger constellation of other ill-defined conditions such as chronic fatigue or irritable bowel syndrome, or is a label given to a variety of physical complaints that arise out of various mental states, such as anxiety and depression. There haven't been any reproducible and clear-cut objective findings, such as blood and lab tests, X-rays, or anatomical abnormalities on biopsy, to provide a satisfactory understanding of the disease. The 1990 American College of Rheumatology diagnostic criteria—widespread muscle pain of more than three months, unassociated with other known illnesses, and the presence of at least eleven tender points over eighteen muscle groups—are nothing more than subjective patient descriptions. (I am not implying that patients with fibromyalgia don't suffer the pains and discomforts that they describe. My concern is the prevailing idea that an fMRI can distinguish between "psychological states" and disease-generated pain.)

In 2002 Georgetown University researchers Richard

Gracely, Ph.D., and Daniel Clauw, M.D., compared how sixteen women with fibromyalgia and sixteen pain-free control subjects responded to both painful and nonpainful stimuli (a small piston applying various amounts of pressure to the base of the thumbnail). They found that control subjects required more than twice the amount of pressure to elicit the same degree of pain and functional image activation as fibromyalgia patients. The authors wrote: "These results convinced us that some pathologic process is making these patients more sensitive. For some reason, still unknown, there's a neurobiological amplification of their pain signals."[24]

Of course there's an amplification of their pain appreciation; otherwise all the subjects would experience the same degree of pain in response to a given stimulus. The real question is whether or not this difference in pain sensitivity represents an underlying disease or simply differences in perception and expectation. To think through how brain activation is related to pain perception, consider how a placebo works. If you believe that an inert sugar pill (placebo) is a powerful analgesic, it can significantly reduce your level of pain from, say, a dental procedure or wear-and-tear arthritis. Conversely, if you are given the same sugar pill and told it is a new untested drug and might make your pain worse, you might experience more pain (nocebo effect). Your expectation of what the pill might do will affect both your pain perception and your fMRI. Nowhere in this schema is there any suggestion that changes in pain perception arising out of your imagination aren't real. Placebo-induced relief of pain is clinically identical to pain relief from standard analgesics such as morphine but tells us nothing about the nature of the pain. Certainly it does not tell us whether the pain is due to "real" or "imagined" causes.

Now consider one of the central features of fibromyalgia—an increased number of areas sensitive to ordinary pressure. If you believe (and are told by your treating physician) that you have a condition that makes you more sensitive to painful stimuli, you are more likely to experience a greater degree of pain than someone who doesn't believe that he is particularly sensitive to painful stimuli. This difference in the level of pain appreciation or description, and the attendant brain changes on fMRI, will be a reflection of your self-perception, not evidence of the presence or absence of disease. Your belief will have the same effect as a nocebo. Even personality traits such as optimism or pessimism (half-empty versus half-full), or one's attitudes toward the medical establishment, can make critical differences.

In a subsequent study, the researchers found a single region of altered activity between fibromyalgia patients and control subjects—in the right thalamus. The magnitude of this difference correlated with the degree of fibromyalgia symptoms; the greater the difference, the worse the patient's symptoms were likely to be. The authors speculated that the findings "are likely the result of neuronal dysfunction."

But these results may also be a reflection of expectation. Psychological profiles have shown that those fibromyalgia patients who believe their pain is the result of some external factor, such as a prior injury or exposure to toxic chemicals, experience a higher degree of altered activity on imaging studies. This belief is also associated with a higher depression rating on the study questionnaire.

An alternative interpretation of the study is that certain areas of brain are activated with expectation of increased pain rather than being the primary cause of this increased pain. And, more important, there is nothing on the scan to point to whether this activity is or isn't "normal." Nevertheless, the authors conclude, "There is really something wrong going on in the brains of the patients with fibromyalgia." According to

Dr. Clauw, "Pain is always a subjective matter, but everything that we can measure about the pain in fibromyalgia shows that it is real."[25]

As a result of these studies, Pfizer was able to argue to the FDA that fibromyalgia is a "real" condition. In 2007, the FDA approved the use of Lyrica for the treatment of fibromyalgia. (Since the approval, it is estimated that worldwide sales of Lyrica have more than doubled, to well over $3 billion annually [as of 2011].)[26] To get a feeling for the circularity of the argument, note that Lyrica has been approved in Europe for the treatment of generalized anxiety, having been shown to be effective in providing relief of both emotional symptoms, such as depressive symptoms and panic, as well as physical symptoms, including headaches and muscle aches.[27]

Will or Intention

Though I am tempted to conclude this chapter with some further observations about free will, I realize that this is a fool's mission. In the larger picture, personal responsibility isn't about free will, but about intention. Did he/she/I consciously or unconsciously intend to do such-and-such, and how do we parse this difference in terms of culpability/responsibility? Whether or not the person experiences agency and conscious choice is beside the point. Harvard psychologist Daniel Wegner adroitly summarizes the problem: "The experience of consciously willing an action is not a direct indication that the conscious thought has caused the action."[28] Rather than focus on free will, we need to direct our attention toward the idea of intention.

If I want to write a novel and try my hardest to think of a good opening, I will experience a sense of effort and a feeling of making choices. If I can't come up with a good introduction and put the project on hold, my intention to write a novel has not been abandoned. "On hold" means that the intention has been transferred to the unconscious, where I will continue to work on the project out of sight. To suggest that the unconscious is willfully trying to solve the problem raises a thorny problem of what the phrase "unconscious willfulness" means. But I think that we all understand that this unconscious rumination is intentional in the sense that it has an intended goal and purpose.

In 1983, Benjamin Libet, a University of California, San Francisco, neurophysiologist, demonstrated the presence of consistent brain activity in the motor region controlling finger movement *prior to* the subject reporting any conscious awareness of the intent to move a finger. Other experiments have confirmed this finding, leading to the belief that the unconscious intention to move the finger precedes any conscious awareness of the intention. There have been a number of potential criticisms of the study, but it remains a seminal paper on the nature of the conscious versus unconscious origin of our decisions.[29] In a more recent update of the Libet experiment, neuroscientist John-Dylan Haynes found brain activity up to ten seconds before a conscious decision to move.[30] His conclusion: "The conscious mind is not free. What we think of as 'free will' is actually found in the subconscious."[31] I prefer interpreting such studies as being evidence for unconscious intention rather than free will, as "unconscious free will" sounds like an oxymoron.

If the primary purpose of debating freedom of choice is to understand and assign personal responsibility, we would be better served by investigating the nature of intention. But, as we've seen, intention is a dynamic interaction between conscious and unconscious brain activity, both past and present. There is no clear demarcation. The most obvious intentional act—premeditated murder—is fueled by unconscious urges and desires. But it is also associated with a high degree of conscious long-term intentionality.

At the other end of the spectrum are those tragic children with Lesch-Nyhan syndrome who bite off their own fingers to satisfy unconscious urges rather than conscious desires. The act is still intentional (as opposed to random or accidental), but at an unconscious level. Another example would be the scatological utterances (coprolalia) of some patients with Tourette's syndrome. But even here, patients will indicate that they have some partial temporary ability to restrain the outbursts.

And what are we to make of addiction? Surely addiction is a function of biology at multiple levels, from the pharmacological effect of a drug or alcohol to the biologically mediated aspects of personality that contribute to the ability to recognize and address the problem. And yet there is considerable neuroscientific literature supporting the mitigating role of personal effort in overcoming addition.[32]

Ultimately, how we conceive of personal responsibility isn't primarily a question of whether or not we have "free will." What are needed are better ways to think about conscious versus unconscious intention—a profound challenge given that unconscious intention is conceptually beyond scientific inquiry.

13 • Tell Me a Story

If others examined themselves attentively, as I do, they would find themselves, as I do, full of inanity and nonsense. Get rid of it I cannot without getting rid of myself. We are all steeped in it, one as much as another; but those who are aware of it are a little better off—though I don't know.

—Michel de Montaigne, "Of Vanity"

My first clinical exposure to neurology, nearly fifty years ago, was watching a senior neurologist examine a forty-five-year-old accountant who, following open-heart surgery, had lost his peripheral vision. All that remained was a slight degree of central vision—the equivalent of looking at the world through two pinholes. Before the neurologist entered the examining room, he had explained that the patient had been acutely paranoid since awakening from surgery. With a wry smile he added that the consulting psychiatrist had diagnosed a postoperative psychosis.

When the neurologist began demonstrating the visual field exam, the patient retreated to the far corner of the room; soon he was standing pressed directly against the side wall of the examining room, bracing himself with his outstretched palms. His look was one of stark terror. Under the gentle coaxing of the neurologist, he said, "I have no sense of my surroundings. Somebody could be sneaking up behind me."

Later the neurologist explained that the accountant had sustained an embolus (blood clot) to his visual cortex, leaving the patient with only pinpoint central vision. After teaching us about the neurology of vision, he shifted gears, becoming pensive. He tentatively suggested that the patient had lost his "mind's eye," and that not having a sense of what was going on outside of his markedly constricted field of vision was making him paranoid.

I still remember being struck by the extraordinary opportunity that neurology offered; you could use scientific knowledge to speculate on what the mind might be. The early clinical case studies of Oliver Sacks give you a feeling for the sense of wonder and mystery that accompanied this period in the history of neurology.

Since that time, the unraveling of the anatomy and physiology of the visual cortex has provided a model for how brain functions are hierarchically organized. We have a much better idea of the mechanisms underlying higher-level visual disorders, such as losing your mind's eye or mistaking your wife for a hat. Overall, the advances in neuroscience have been spectacular, the result of great innovative thought and collaboration. I am in utter awe of the brilliance that has brought neuroscience from the dark ages to the present level of sophistication regarding brain function.

But how we ultimately think about the accountant who became paranoid after losing his peripheral vision is more than scientific explanation. Each of us brings to his or her observations an entire worldview created by both biology and experience. And though this might seem like a disquieting, if not downright threatening, conclusion, this set of prejudices goes for neuroscientists too.

Neuroscientists must acknowledge that translations of scientific data into causal explanations about the mind are pure storytelling. This is not to discredit the science of the brain any more than it is to discredit a good forensics investigation of a murder scene. But if the evidence is strictly circumstantial—be it an unwitnessed crime or a subjective mental state—we must acknowledge where data ends and story begins.

Neuroscientists are like mystery writers. In challenging us to solve a whodunit, a writer plants clues; a neuroscientist offers data. The data may be achieved via scientific method, but the simplest tale—the stroke occurred and the accountant became fearful—is the description of a sequence of events dependent upon everything from the neuroscientist's innate sense of causation to his own experience with inexplicable fear and trembling.

Studying the mind isn't like other areas of science, where accurate measurements can be made without significant intrusion of perceptual biases. A physicist can measure the speed of light without great concern that his measurements will be influenced by his politics, religious beliefs, or innate predispositions, or by the smell of bakery goods. The same is not true in neurology. There are no mind measurements; there are only stories derived from scientific data and filtered through personal perceptions.

The history of science is the back-and-forth movement of trial-and-error advances and retreats, punctuated by moments of brilliance and marred by periods of excess. Today I fear that neuroscience is teetering on the brink of an era of excess. If we persist in arguing that political candidates can be judged by the relative activity of their amygdala or anterior cingulate gyrus, or that a lowered libido can be measured by an fMRI, we can be sure that history won't treat this era of neuroscience kindly.

To see how important it is to read neuroscientific observations as a story told by a narrator with his or her own inbuilt biases, let me present one final series of ethically loaded case studies by a single author. The question I'd like to pose is whether or not these studies would be easier to judge if we knew more about the author.

One of the most difficult issues in dealing with patients with profoundly impaired cognitive function such as PVS is whether or not to withdraw life-sustaining care—so-called passive euthanasia. (When physicians talk about euthanasia in the severely neurologically compromised, they are referring to the passive form, where nutrition and fluids are withheld, not the active form, where a patient is deliberately administered lethal medications.) The ethical issues are numerous; there is no clear-cut right answer. To make the optimal decision, family members (often with diametrically opposed points of view) must rely upon the best available medical evidence regarding the accuracy of diagnosis, the likelihood of a meaningful recovery, and the reasonableness of new or alternative therapies. Ideally this information should be presented free of bias on the part of the researchers. Unfortunately, it is hard to imagine any researcher without some feelings on an issue as emotionally charged as deciding whether or not to let a patient die. Yet these biases (whether conscious agenda or subliminal feelings unrecognized by the researcher) inform every aspect of the study, from the reasons for and the design of an investigation, to the choice of methodology, to statistical analysis, and ultimately, to how to interpret the data. The greater the stakes and emotional pull of a particular point of view, the more likely the final interpretation will be a reflection of these biases. In reading the following series of articles, imagine how differently they might be interpreted if we knew the author's personal background (including his religious/nonreligious beliefs and feelings about euthanasia).

Locked In

2009 headline: "Man Says Emergence from 'Coma' Like Rebirth."[1]

In 1983, a young man, RH, was involved in a car accident that left him in a presumed persistent vegetative state. Twenty-three years later he was referred to Steven Laureys, M.D., Ph.D., of the University of Liège in Belgium, a specialist in disorders of consciousness. After an fMRI examination, Dr. Laureys reported that RH's prior doctors were wrong and that RH was in a minimally conscious state. The family hired a speech therapist to help RH communicate via a touch-screen computer. Three years later, the Associated Press reported that RH was now conscious and, using this method, was able to fully communicate. The speech therapist, according to the Associated Press, said that she could feel RH "guiding her hand with gentle pressure from his fingers, and that she feels him objecting when she moves his hand toward an incorrect letter." With her help, she said that RH typed: "It was especially frustrating when my family needed me. I could not share in their sorrow. We could not give each other support. Just imagine, you hear, see, feel and think that no one can see that. You undergo things. You cannot participate in life."

Arthur Caplan, a bioethics professor at the University of Pennsylvania, was skeptical of the technique of the therapist interpreting the patient's finger movements. He described it as "facilitated communication . . . Ouija board stuff that has been discredited time and time again." Caplan also suspected that the patient's statements were inconsistent with such a profound injury and an inability to communicate for decades.

When interviewed, Dr. Laureys indicated that each year he evaluated about fifty such patients from around the world, and that he was in the process of reexamining dozens of them. He made no comments about the criticism of RH's method of communication, which the speech therapist had devised.

Later, after RH's mother indicated that her son was writing a book about his experience, Laureys conducted a simple test to assess the validity of the patient's communication. The speech therapist was asked to leave the room. RH was shown various objects and later asked to type their names with assistance from a neutral observer. Despite repeated attempts, RH was unable to identify any of the objects or provide any meaningful communication. When asked why he hadn't expressed any earlier skepticism about the therapist's role in RH's newfound ability to communicate, Laureys replied, "The story of RH is about the diagnosis of consciousness not communication. From the start I did not prescribe this technique. But it is important not to make judgments. His family and caregivers acted out of love and compassion."[2]

One point about methodology: There are standard methods for detecting minimal motor movements. By placing electrodes over the muscles to the finger, even movements that are too small to see can be detected electrically and displayed on a monitor. In the same way as "locked-in" patients (who are fully conscious but are paralyzed and cannot speak) can communicate with any remaining movement such as an eyeblink, as was the case of magazine editor Jean-Dominique Bauby, whose story was immortalized in the book and the movie *The Diving Bell and the Butterfly*, it would have been possible to work out an arrangement whereby RH could have directly communicated by viewing the monitor. There was no need to introduce an intermediary to translate the communication.

In 2011, two years after his initial comments about RH, Laureys published a study in the *British Medical Journal* assessing the quality of life of sixty-five patients with locked-in syndrome. The patients answered a series of questions by blinking. Forty-seven said they were happy and eighteen indicated that they were unhappy. Laureys wrote that the findings should change not only how patients are treated, but people's attitudes toward euthanasia. He was modestly optimistic that with rehabilitation, many patients could regain some control of their head, fingers, and feet and might be able to talk a little.

His overall conclusion: "Our data stress the need for extra palliative efforts directed at mobility and recreational activities in locked-in syndrome. . . . Recently affected locked-in syndrome patients who wish to die should be assured that there is a high chance they will regain a happy meaningful life. . . . Patients' requests for euthanasia should be received with sympathy, but our data suggest that a moratorium should be proposed until the patients' condition has stabilized." Laureys and colleagues noted that the longer a person had locked-in syndrome, the more likely he or she was to be content.[3]

To propose a moratorium on euthanasia because of an undocumented belief that better rehabilitation techniques might provide a better quality of life is to put the moral cart before the scientific horse. To date, there is scant evidence about major benefits of long-term rehabilitation in patients with profound disorders of consciousness or locked-in syndrome. Equally bothersome is how to determine that a patient has stabilized when Laureys tells us that the longer a patient waits, the more likely he is to be content with his situation. One of the conditions for considering withdrawal of treatment is the high likelihood of no further recovery; if you argue (without adequate substantiation) that a person's emotional state is likely to improve with the passage of time, withdrawal of support can never be a realistic option. In addition, how reliable do you think the patient's description of his mental state is likely to be when his care is totally dependent upon those asking the question? Imagine the personal nightmare of being nearly totally paralyzed and trying to figure out how to answer honestly and yet not offend your caretakers. And what are we to make of Laureys advising other physicians to assure their similarly afflicted patients that they are quite likely to have a happy meaningful life?

The underlying data of Laureys's studies can be judged on their scientific accuracy. That is as far as science can go in determining the validity of his observations. But his interpretations of his results aren't science; they are cautionary tales about our clinical inability to accurately determine levels of consciousness and why we should be leery of euthanasia in such patients. As neuroscience often does not present us with clear-cut distinctions between data and story, it is up to us to draw the line.

Let's return to the 2007 *Archives of Neurology* article, coauthored by Dr. Laureys, indicating that the young woman, patient X, was conscious because her fMRI showed her imagining playing tennis and walking around her house. Now that we know Dr. Laureys's position on the relative happiness of patients with locked-in syndrome and his concerns about euthanasia, should we not question whether his beliefs have any bearing on his construction and interpretation of studies of patients with other disorders of consciousness, especially when his conclusions have such far-reaching implications?

It would be easy to provide alternative interpretations of Laureys's studies on happiness, or to condemn his uncritical acceptance of a discredited methodology (facilitated communication). But personal criticism of individual studies isn't going to highlight the larger issues: we must acknowledge the limits of what science can say about another's state of mind and understand the moral consequences of representing personal visions as scientific fact.

I have to admit I'm bothered by my own criticism of the research of Laureys and colleagues, because it has yielded valuable techniques for improving our understanding of basic brain function in altered states of consciousness. The methods are ingenious and provocative; they are already triggering other fascinating research. But even the most outstanding and meticulous research shouldn't be a license to express moral conclusions as though they are unassailable scientific facts. If personal views of intelligence, consciousness, or morality are offered as scientific truths, such neuroscience is no better than the faith-driven absolutes peddled by the opponents of scientific method.

Wrapping Up

When reading any neuroscientific claim about the mind, remember:

- All thoughts about and studies of the mind are guided by involuntary brain mechanisms that collectively generate an illusory sense of a personal, unique self capable of willful, unbiased exploration of how a brain creates a mind.

- Taking into consideration how these involuntary mental states create our sense of a mind is a necessary first step to any real, albeit partial, understanding of what a mind might be.

- Failure to acknowledge the biologically imposed limits on a mind examining itself will only result in further neuroscientific excesses.

Wisdom Is Humility in the Face of Mystery

If I were asked to distill this book into a single message, I would say that all of us—neuroscientists, cognitive scientists, psychologists, philosophers, and lay readers—should be constantly aware of the essential paradox that drives all investigations of the mind. The mind exists in two different dimensions—as felt experience and as abstract concept. The unavoidable takeaway is that a constellation of involuntary mental sensations play a critical role in how we think about what a mind "is" and "does." It is the human condition to experience a largely involuntarily generated mind that feels quite strongly that it can rationally explain itself. This paradox is unavoidable and not amenable to better science or new technologies. Though we can and should work hard to refine our thoughts, there will always be limits. Ironically, even if there were to be a final and absolute last word on the nature of the mind, we wouldn't recognize it unless we all thought the same way—and that's a physiological impossibility.

In their study "Unskilled and Unaware of It," which I detailed in chapter 5, Justin Kruger and David Dunning provide one of the best conclusions to a neuroscientific article that I have read. I offer it as a model of what good science of the mind can and cannot say about itself.

"Although we feel we have done a competent job in making a strong case for this analysis, studying it empirically, and drawing out relevant implications, our thesis leaves us with one haunting worry that we cannot vanquish. That worry is that this article may contain faulty logic, methodological errors, or poor communication. Let us assure our readers that to the extent this article is imperfect, it is not a sin we have committed knowingly."[4]

This article summary is honest, recognizes its own

inherent limitations, isn't offered as the last word, and is presented with style, wit, and humility. Judging by the character of the conclusion, the authors have integrity and are leveling with their readers.

To scientists, the summary offers a model for the future. Not one of us, not the smartest or cleverest or most profound neuroscientist, philosopher, or observer of mankind, has the final word. Each of us is weaving stories, not uncovering absolute truths. The mind is and will always be a mystery. For the neuroscientist, a humble acknowledgment of the limits of inquiry should be the first step in the study of the mind. If this means that neuroscientists must step outside their circles of training and personal confidence in order to consider how their personal selves are subliminally driving their conclusions, so be it. To continue with the untenable pretense that our understanding of the mind requires only irrefutable data is to ignore what we are learning about how the brain works.

The job for the lay reader is a bit different. Few of us have a sufficient scientific background to fully assess the raw data of neurology. Yet we all have a sense of what is a good or a bad story. With a work of fiction, we often consider the relationship of the author to the story. We look at the jacket blurbs and check out the author's photograph. We check out his website to get some biographical information, descriptions of previous books, and perhaps a short statement on why he wrote the book. We take it for granted that knowledge of the author will enhance our understanding and appreciation of the story he's telling.

Readers should take the same approach with neuroscientists. And neuroscientists should be required to oblige them. What we need from scientists is an understanding of how and why they have chosen their particular subject matter, methodologies, and interpretations. We need some idea of what makes the author of each study tick. Though a public self-revelation for a scientist runs counter to the traditional but unwarranted presumption that science is entirely objective and should be stripped of personal statements, our understanding of neuroscience would be very different if every study contained a paragraph or two in which the author revealed his or her understanding of what personal motives and agendas drove the study. No matter how faulty or incomplete self-knowledge is, some glimmer of an author's underlying motivations and urges would be invaluable. At the very least, this additional information would allow each of us to judge the veracity and consistency of the author's stated reasons for the study, detect possible hidden agendas, consider the level of the author's own "self-awareness," and get some sense of the author's "character" that is telling the story.[5]

Neuroscientists are quickly becoming the preeminent narrators of the modern story of the mind. They have the tools, language, and experience to tell us informed, engaging, and important stories. In turn, we should judge their studies in the same light in which we judge other artistic forms. We should assess the preciseness of language, the tightness of structure, the clarity and originality of vision, the overall elegance and grace of the study, the restraint with which they present moral issues, place their studies in historical, cultural, and personal context, and their willingness to seriously entertain alternative opinions and interpretations. Just as a good novelist acknowledges that his portrayal of a character, no matter how wonderful, isn't necessarily the only way to depict the character, neuroscientists must see their conclusions about the mind as one interpretation among several or many. After all, all conclusions about the mind are personal visions, not irrefutable and inevitable consequences of scientific thought.

Great art is the expression of reverence and wonder. It is

also the investigation of and acknowledgment of limits. The operating credo of neuroscience should be a rigorous adherence to scientific method coupled with the recognition that the study of the mind is a data-based art form, not another branch of the basic sciences. Humility, reverence, and respect for the unknowable should be the default mind-set when contemplating great mystery, and there is nothing more mysterious than a mind contemplating itself.

Acknowledgments

Acknowledging those who have inspired an idea isn't easy. As this book is the product of long-standing rumination rather than detailed scientific research, there hasn't been a discrete starting point, nor any specific milestones or even a clear path. Though many sources of inspiration have been obscured by time and the ravages of memory, what remains is their collective subliminal impression, along with the feeling of gratitude for the enormous privilege and adventure that my career in neurology has offered me. My thoughts about the mind have been shaped by fellow residents, colleagues, students, mentors, and some of the great figures in neurology. But first and foremost, they have roots in the observation of and conversation with patients. It sounds cruel to thank people for having suffered often tragic neurological deficits. However, without these brave and often keenly observant patients who have been generally tolerant of prodding questions and tests, there would be no neurology of the mind.

As this book is an outgrowth of my previous book, *On Being Certain,* thanks overlap. I am particularly grateful for the guidance, keen eye, and personal touch of my agent, Andrew Stuart. Once again, my editor, Nichole Argyres, has provided ongoing enthusiastic support and has insightfully improved on my original manuscript. Her assistant, Laura Chasen, has rounded out an unsurpassed editorial dream team. Fortunately, my friends remain my friends; particular thanks go to those who've born the brunt of my often tangential if not outright obscure enthusiasms. At the top of the list are Kevin Berger, John Campbell, David DiSalvo, David Dobbs, Jonathon Keats, Barbara Oakley, Peter Robinson, David Rubin, and Richard

Segal. The members of the San Francisco Philosophy Club have provided both sanctuary and provocation. The Institute for the Future has been a great sounding board. The research librarians at Mt. Zion-UCSF library—Gail Sorrough, Gloria Won, and John Phillips—have been invaluable. Thanks also to the many readers of *On Being Certain* who took the time to send their often challenging and provocative comments, many of which have prompted aspects of the present book.

As always, I reserve the deepest thanks for my family. My parents, always practical, nevertheless encouraged (or at least put up with) what would become my life of impractical musings. My wife, Adrianne, has been the great inspiration of my life. To the degree that you enjoy this book, thank her for her unfailing wisdom and level-headed critiques. When you disagree with my ideas, blame me for not listening to her.

Notes

Introduction

1. http://booklover.tumblr.com/post/359461486/a-book-is-the-only-place-in
-which-you-can-examine.

2. Given these harsh realities, it isn't realistic to demand that neuroscientists be
held to a higher standard of self-restraint, personal reserve, and sense of the
moral dimension of their opinions than their friends, neighbors, and academic colleagues in other fields. There is no neuroscientific equivalent of the
Hippocratic oath: Above all, do no harm.
    For the foreseeable future, the study of the mind will continue to have a
lawless Wild West quality. I am tempted to suggest the analogy to a flea market devoted to selling models of a mind. Everyone is hawking his personal
contraption. Some will be utterly worthless, some will be of marginal value,
some will be wonderful at first but break with continued use, while others
will be exactly as advertised. And there will be true gems for sale. As there is
no governing body to police the information, no neuroscientific *Consumer
Reports* to test the products, no Better Business Bureau to whom you can report misrepresentation or downright fraud, and no Good Housekeeping Seal
of Approval to reward the best products, it is up to each of us to be a discerning consumer.

3. E. Callaway, "Possible Site of Free Will Found in Brain," *New Scientist*, May
7, 2009, http://www.newscientist.com/article/dn17092-possible-site-of-free
-will-found-in-brain.html?

4. http://www.independent.co.uk/life-style/health-and-families/health-news/bad
-behaviour-down-to-genes-not-poor-parenting-says-study-2093543.html.

5. A. Damasio, "The Brain: A Story We Tell Ourselves," *Time,* January 29, 2007,
http://www.time.com/time/magazine/article/0,9171,1580386,00.html
#ixzz12IU4qXQa. "Some philosophers maintain that solving the problem of
consciousness is beyond the reach of human intelligence. This is very odd and,
I believe, untrue. . . . All the natural history required to understand consciousness is now readily available in evolutionary biology and psychology."

6. M. Brooks, "What We'll Never Know," *New Scientist*, May 7, 2011, 38. In this
article, Brooks states that there are 100 million neurons in the brain, but
there are actually 100 billion.

7. Neurons have a greater metabolic rate when more active, hence require
higher levels of oxygen. By determining the relative levels of blood oxygen
in various areas of brain (BOLD technique or blood oxygen level determination), the fMRI can demonstrate changes in level of brain activation when a
subject is performing a task. As the technique relies on inferring brain states
from measurements of blood flow, there are a number of theoretical and
practical limitations and potential pitfalls. For a detailed discussion of the
limits of fMRI scanning, read N. Logothetis, "What We Can Do and What

We Cannot Do with fMRI," *Nature*, June 12, 2008, 869–78 and/or http://
blogs.scientificamerican.com/guest-blog/2012/07/05/controversial-science
-of-brain-imaging/. To get a sense of the willingness of some scientists to
suspend disbelief about these limits, see: http://blogs.scientificamerican.com
/observations/2012/06/21/whats-a-voxel-and-what-can-it-tell-us-a-primer
-on-fmri/.

1. The Shape of Your Mind

1. http://www.searchquotes.com/search/branchtrickey/.
2. "Llinás argues convincingly that the centralization of motor control produces the need for the organism to monitor and predict its own bodily state.
   The pervasive, intimate, moment-to-moment 'sense of self' that we all enjoy
   is thus to be understood not as the product of some evolutionary leap in
   cognitive or perceptual sophistication, but as a functional prerequisite for the
   deliberate control of action." http://www.americanscientist.org/bookshelf
   /pub/from-motricity-to-mentality.
3. P. Brugger, R. Agosti, M. Regard, H. Wieser, and T. Landis, "Heautoscopy,
   Epilepsy, and Suicide," *Journal of Neurology, Neurosurgery and Psychiatry* (1994):
   838–39. The technical name for this condition is heautoscopy; it is most
   commonly seen in patients with parietal or deep-temporal-lobe origination
   of their seizure disorder. This description combines the classic features of
   autoscopy, which is the visual hallucination of one's body or body parts, as in
   a mirror reflection, with an out-of-body experience in which the primary
   illusion is of being separated from one's own body.
4. http://www.scientificamerican.com/article.cfm?id=neuroscience-of-self
   hood.
5. A number of species are now thought to have at least a rudimentary sense of
   self as demonstrated by self-recognition in a mirror. The greater the complexity of behavior, the greater the requirement for a sense of self necessary
   for more complex thoughts and actions.
6. http://scienceblogs.com/neurophilosophy/2008/12/the_bodyswap_illusion
   .php.
7. For a summary of recent brain localization studies, two good references are:
   S. Gallagher, "Multiple Aspects in the Sense of Agency," *New Ideas in Psychology* (2010), doi: 101016/j.newideapsych.2010.03.003 and N. David, A. Newen,
   and K. Vogeley, "The 'Sense of Agency' and Its Underlying Cognitive and
   Neural Mechanisms," *Consciousness and Cognition* 17 (2008): 523–34.
8. "It makes sense that self-awareness should be processed in regions of the
   brain that integrate many different sources of inputs together. Since the different sources of external inputs carry different aspects of information about
   the organism, these respective aspects will all be strongly correlated with
   one another. In an organism with a sufficiently powerful associative memory
   and learning ability, these various self-inputs will result in the formation of a
   highly correlated, multi-dimensional 'super-input.'
       "It should be emphasized that the self-image is not necessarily a well-
   defined set of neural pathways that represents a specific internal picture of

the organism. Rather, it should be thought more of as a set of associations generated by the continual input of information related to the organism itself. While such associations may provide a nucleus of pathways that form the initial self-image, the formation of memories involving the self-image may themselves become part of the self-image, so that the number of associations defining the self-image can increase in time." E. Tannenbaum, "Speculations on the Emergence of Self-Awareness in Big-Brained Organisms," *Consciousness and Cognition* 18 (2009): 414–27.

9. http://www.shoreline.edu/psparks/homework/OBE.pdf.

10. Neurologists often use the term "disconnected" or "disconnection syndrome" to refer to the uncoupling of two aspects of an act that normally occur as a single action. For example, writing a sentence and reading what you have written are normally experienced as a unit. However, with localized brain injuries such as a stroke, one function can be selectively impaired while another remains intact—resulting in such peculiar syndromes as the ability to write a sentence yet not be able to read what you've written (alexia without agraphia).

Blanke and Metzinger have extended their view of these basic feelings of self by categorizing them as composing a "minimal phenomenal selfhood." "Phenomena" refers to those experiences that give us a sense of the presence of a self. "Minimal" means that we're excluding the more complex narrative sense of self created by memories, experiences, stories, etc. Various researchers offer somewhat different categories, but most concur that there are several separate experiences that collectively create this minimal selfhood. These include a sense of ownership and identification with the body as a whole (a sense of "mineness"), where the self is located in space, and the orientation from which we see the world—the first-person perspective. Collectively these constitute "the experience of being a distinct, holistic entity capable global self-control and attention, possessing a body and a location in space and time." O. Blanke and T. Metzinger, "Full-Body Illusions and Minimal Phenomenal Selfhood," *Trends in Cognitive Sciences* 13, no. 1 (2008): 7–12.

11. For a video demonstration: http://www.youtube.com/watch?gl=US&hl=uk &v=TCQbygjG0RU.

12. http://www.plosone.org/article/info:doi/10.1371/journal.pone.0003832; V. I. Petkova and H. H. Ehrsson, "If I Were You: Perceptual Illusion of Body Swapping," *PLoS One* 3 (12): e3832. doi:10.1371/journal.pone.0003832. *PLoS* is the Public Library of Science.

13. Youtube demonstration: http://www.youtube.com/watch?v=rawY2VzN4-c.

14. A. Iriki, M. Tanaka, and Y. Iwamura, "Coding of Modified Body Schema During Tool Use by Macaque Post-central Neurons," *Neuroreport* 7, no. 14 (1996): 2325–30.

15. S. Obayashi, T. Suhara, K. Kawabe, T. Okauchi, J. Maeda, Y. Oakine, H. Onoe, and A. Iriaki, "Functional Brain Mapping of Monkey Tool Use, *Neuroimage* 14, no. 4. (2001): 853–61.

16. M. Quallo, C. Price, K. Ueno, T. Asamizuya, K. Cheng, R. Lemon, and A. Iriki, "Gray and White Matter Changes Associated with Tool-Use Learning in Macaque Monkeys," *Proceedings of the National Academy of Science* 106, no. 43, October 27, 2009: 18379–84, http://www.pnas.org/content/106/43/18379 .full.pdf+html.

17. E. Gould, "How Widespread Is Adult Neurogenesis in Mammals?," *Nature Reviews Neuroscience* 8 (2007): 481–88.

18. R. Peeters, L. Simone, K. Nelissen, M. Fabbri-Destro, W. Vanduffel, G. Rizzolatti, and G. Orban, "The Representation of Tool Use in Humans and Monkeys: Common and Uniquely Human Features," *Journal of Neuroscience* 29, no. 37, September 16, 2009: 11523–39, http://www.jneurosci.org/content /29/37/11523.full.pdf.

19. G. Berlucchi and S. Aglioti, "The Body in the Brain: Neural Bases of Corporeal Awareness," *Trends in Neuroscience*, 20 (1997): 560–64.

20. Philosophy 132, UC Berkeley Lecture Series by John Searle on iTunes U, Spring 2010.

21. In the 1930s, H. Kluver and P. Bucy showed strikingly diminished fear responses in primates with damage to the region of the temporal lobe where the amygdala is located. A more recent study: M. Meunier, J. Bachevalier, E. A. Murray, L. Málková, and M. Mishkin, "Effects of Aspiration vs. Neurotoxic Lesions of the Amygdala on Emotional Responses in Monkeys," *European Journal of Neuroscience* 11 (1999): 4403–18.

22. Lipoid proteinosis or Urbach–Wiethe disease.

23. http://www.nytimes.com/2010/12/21/science/21obbrain.html.

24. D. Kennedy, J. Glascher, J. Tyszja, and R. Adolphs, "Personal Space Regulation by the Human Amygdala, *Nature Neuroscience* 12, no. 10, October 2009: 1226–27.

25. A. Järvinen-Pasley, U. Bellugi, J. Reilly, D. Mills, A. Galaburda, R. Reiss, and J. Korenberg, "Defining the Social Phenotype in Williams Syndrome: A Model for Linking Gene, the Brain, and Behavior," *Development and Psychopathology* 20 (2008): 1–35.

26. http://www.jstor.org/pss/2786318 J. Baxter, "Interpersonal Spacing in Natural Settings," *Sociometry* 33, no. 4 (1970): 444–56.

27. http://www.foxnews.com/story/0,2933,520811,00.html.

28. P. McGeoch, D. Brang, T. Song, R. Lee, M. Huang, and V. Ramachandran, "Apotemnophilia—the Neurological Basis of a 'Psychological' Disorder," *Nature Precedings*, hdl:10101/npre.2009.2954.1, http://precedings.nature.com/ documents/2954/version/1/files/npre20092954-1.pdf.

29. V. Ramachandran, D. C. Rogers-Ramachandran, and S. Cobb, "Touching the Phantom," *Nature* 377 (1995): 489–90.

30. K. MacIver, D. M. Lloyd, S. Kelly, N. Roberts, and T. Nurmikko, "Phantom Limb Pain, Cortical Reorganization and the Therapeutic Effect of Mental Imagery," *Brain* 131, no. 8 (2008): 2181–91, http://brain.oxfordjournals.org /content/131/8/2181.short.

31. R. Thompson, "Amputees could get a helping hand in the virtual world," *New Scientist,* March 16, 2010.

32. M. Costandi, "Distorted Body Images: A Quick and Easy Way to Reduce Pain," *Scientific American*, December 23, 2008, http://www.scientificamerican .com/article.cfm?id=a-quick-way-to-reduce-pain.

33. For V. S. Ramachandran's review of mirror box therapy in a variety of disorders, see http://brain.oxfordjournals.org/content/early/2009/06/08/brain .awp135.full.

2. Agency, Will, and Intention

1. I. Farber, "The World Within the Skull," *Nature,* 413, 776–77.

2. In *On Being Certain,* I suggest that the sense of purpose is best seen as an involuntary mental sensation closely related to the feeling of knowing. See chap. 13, pp. 177–87.

3. I don't mean to imply that there is actually a specific brain region that can be considered the site of predictions. Rather, predictions are a function of the entire neural circuitry. However, the concept of a central predictor is a handy combination metaphor and shorthand.

4. G. Lafargue and N. Franck, "Effort Awareness and Sense of Volition in Schizophrenia," *Consciousness and Cognition* 18 (2009): 277–89.

5. G. H. Lewes, "Motor-Feelings and the Muscular Sense," *Brain* 1 (1878): 14–28. In 1805 the French philosopher Maine de Biran wrote that the sense of effort is the fundamental component of self-experience, the landmark of the exercise of the will.

6. Similar findings have been observed in volunteers who have been subjected to experimentally induced peripheral sensory loss.

7. C. Frith, *Making Up the Mind: How the Brain Creates Our Mental World* (Oxford, UK: Blackwell Publishing, 2007), 103.

8. S. J. Blackmore, D. M. Wolpert, and C. Frith, "Central Cancellation of Self-Produced Tickle Sensation," *Nature Neuroscience* 1, no. 7 (1990): 635–40.

9. C. Frith, "The Self in Action: Lessons from Delusions of Control," *Consciousness and Cognition* 14, (2005): 763.

10. To get a general sense of the difficulty of assigning a complex mental sensation to a single area or even a closely related neural system, see N. David, A. Newen, and K. Vogeley, "The 'Sense of Agency' and Its Underlying Cognitive and Neural Mechanisms," *Consciousness and Cognition* 17 (2008): 523–34.

11. *Making Up the Mind,* 105.

12. http://www.youtube.com/watch?v=59YKlP—PhU&NR=1.

13. http://www.thepsychologist.org.uk/archive/archive_home.cfm/volumeID_18-editionID_128-ArticleID_931-getfile_getPDF/thepsychologist/1005dell.pdf.

14. In reviewing thirty-nine detailed cases Della Sala et al. (1994) found that most of the patients showing anarchic hand had a lesion encroaching upon the medial wall of the frontal lobe contralateral to the wayward hand. In particular, lesions seem to be centered on an area known as the supplementary motor area (SMA). Each SMA lies in the medial surface of one of the frontal lobes. It plays a role in the execution of movements. It is thought to be responsible for converting intention into self-initiated actions, or to be involved in the selection of what movement to make. Several studies converge in demonstrating that the part of the SMA known as the SMA-proper stores and organizes motor subroutines related to internal drives. S. Della Sala, C. Marchetti, and H. Spinnler, "The Anarchic Hand: A Frontomedial Sign," in F. Boller and J. Grafman, eds., *Handbook of Neuropsychology* (Amsterdam: Elsevier, 1994), vol. 9, 233–55.

15. http://www.thepsychologist.org.uk/archive/archive_home.cfm/volumeID
_18-editionID_128-ArticleID_931-getfile_getPDF/thepsychologist
/1005dell.pdf.

16. G. Banks, P. Short, A. J. Martinez, R. Latchaw, G. Ratcliff, and F. Boller, "The Alien Hand Syndrome: Clinical and Postmortem Findings," *Archives of Neurology* 46 (1989): 456–59.

17. Anarchic hand is often referred to in the literature as "alien hand." However, "alien hand" is a more general term, with different meanings assigned by different authors. According to Della Sala, the confusion arose owing to a mistranslation from the French and dragged on in subsequent scientific reports. For a full account see C. Marchetti and S. Della Sala, "Disentangling the Alien and Anarchic Hand," *Cognitive Neuropsychiatry* 3 (1998): 191–207.

18. This list is by no means complete; there are more complex mental sensations that are harder to pin down. Though some may turn out to be discrete brain functions, most are best seen as the end result of the mixing and matching of more basic mental sensations.

3. Causation

1. A change in the spatial configuration of the two events can also alter any sense of causation. If you stub your toe but a couple of days later develop shin, knee, thigh, hip, or elbow pain, you will assign different likelihoods that the pain is due to the injury depending on how far away from the injury the pain is located.

2. http://news.yahoo.com/s/nm/20100902/lf_nm_life/us_britain_hawking/print.

3. http://www.independent.co.uk/life-style/health-and-families/health-news/bad-behaviour-down-to-genes-not-poor-parenting-says-study-2093543.html.

4. http://www.guardian.co.uk/society/2010/oct/03/oliver-james-adhd-attention-deficit.

4. Sensational Reason

1. Rather than clutter up the discussion with repeated references to "subconscious or unconscious cognitive processing"—clumsy phrase—I'll often substitute in the shortcut word "brain." I realize that this characterization can be misleading, as all mental activity arises from the brain, but it serves to distinguish lower-level mental functions from those believed to represent conscious thought.

2. Though the underlying biology of computation isn't known, we do have the neurochemical precedent of inhibitory and excitatory neurotransmitters that collectively determine whether or not a cell will fire. In the most basic sense, this is a complex calculation of all the weightings within the hidden layer.

3. C. Chabris and D. Simons, *The Invisible Gorilla: And Other Ways Our Intuitions Deceive Us* (New York: Crown, 2010).

4. R. Adair, *The Physics of Baseball* (New York: Harper Perennial, 2002), 42. For a more complete explanation, see pp. 66–80 of *On Being Certain*.

5. A. Damasio, "Remembering When," *Scientific American,* September 2002, 66.

5. Logic's Reason

1. J. Kruger and D. Dunning, "Unskilled and Unaware of It: How Difficulties in Recognizing One's Own Incompetence Lead to Inflated Self-Assessments," *Journal of Personality and Social Psychology* 77, no. 6 (1999): 1121–34.

2. The authors considered the degree of miscalibration rather than simply taking the percentage over estimation in order to take into account that those with a low score would have a greater likelihood of granting themselves a higher score than someone at the top end of the scoring range. To do this, they compared the participants' estimate of how many questions they got right with the actual test scores.

3. J. Hadamard, *The Psychology of Intuitive Judgment* (Cambridge, UK: Cambridge University Press, 1954).

4. B. Russell, "The Study of Mathematics," in *Mysticism and Logic: And Other Essays* (London: Longman, 1919), 60, http://books.google.com/?id=zwMQ AAAAYAAJ&pg=PA60&dq=Mathematics+rightly+viewed+possesses+ not+only+truth+but+supreme+beauty+a+beauty+cold+and+austere+like +that+of+sculpture+without+appeal+to+any+part+of+our+weaker+ nature+without+the+gorgeous+trappings+inauthor:Russell.

5. K. Devlin, *The Math Gene: How Mathematical Thinking Evolved and Why Numbers Are Like Gossip* (New York: Basic Books, 2001), http://books.google .com/books?id=AJdmfYEaLG4C&pg=PA140&lpg=PA140&dq=Why+are +numbers+beautiful%3F+It's+like+asking+why+is+Beethoven's+Ninth +Symphony+beautiful.+If+you+don't+see+why,+someone+can't+tell+ you.+I+know+numbers+are+beautiful.+If+they+aren't+beautiful,+noth ing+is.&hl=en.

6. S. Topolinski and R. Reber, "Immediate Truth-Temporal Contiguity Between a Cognitive Problem and Its Solution Determines Experienced Veracity of the Solution," *Cognition* 114, no. 1 (January 2010): 117–22, doi:10.1016/j. cognition.2009.09.009.

7. Reber's article cites other processing-speed studies that have been used to explain how we judge the accuracy of aphorisms, why we prefer some brands over others, and why we tend to reject mathematical theories that are difficult to understand.

8. S. Hawking and L. Mlodinow, *The Grand Design* (New York: Bantam, 2010). "Because there is a law such as gravity, the universe can and will create itself from nothing": http://www.guardian.co.uk/science/2010/sep/02/stephen -hawking-big-bang-creator; http://news.yahoo.com/s/nm/20100902/lf_nm _life/us_britain_hawking/print>.

9. To get a sense of how an underlying antitheological message drives a first-rate scientist to redefine the nature of nothing, dismiss classical logic, and assign agency to the universe, consider the comments of Lawrence Krauss, theoretical physicist and author of *A Universe from Nothing: Why There Is*

Something Rather Than Nothing (New York: Free Press, 2012). Krauss's explanation: "I think it is virtually certain that everything we see came from empty space. All the physics I know is highly suggestive that our universe popped into existence as a quantum fluctuation." For Krauss, this "empty" space consisted of a bubbling brew of virtual particles spontaneously popping in and out of existence on timescales too small to notice. This was the "nothing" out of which the universe arose. When a *Skeptic Magazine* interviewer asked how something from nothing is possible, Krauss answered, "I'm not interested in classical, logical descriptions of nothing, but rather what science tells us about nothing. The philosophical bother may be there, but who cares? It just means the universe is cleverer than theologians are. The fact that you can't get your head around it doesn't make it less so." A. Z. Williams, "A Universe from Nothing? A Conversation with L. Krauss," *Skeptic Magazine* (e-Skeptic edition), May 30, 2012.

10. I am reminded of Wittgenstein's example of proper syntax creating a sense of meaning in a nonsensical proposition: "It is afternoon on the sun."

11. Ask the calculator to find the exact determination of π and it will go on calculating indefinitely. If programmed to stop only with a final answer, it has no built-in circuit breaker. We too seem constitutionally programmed to find answers to questions irrespective of whether or not they exist. The origin of the universe will always be a top priority for us, as it seems to offer the key to knowing who we are. No amount of evidence for the limits of our understanding will slow us down.

12. This is no different from the polar bear being the first image your brain generates when it hears the term "climate change," while another sees long lines of cars waiting at gas stations. As we've all experienced, our opinions are strongly influenced by these initial images.

13. Hawking emphasizes that perspective is critical to understanding the origin of the universe. He makes a clear distinction between a worm's-eye view from inside the universe and an angel's-eye view from outside the universe. That he claims that his theory is predicated upon an angel's-eye view is yet another example of how the mind's-eye perspective is at the heart of cosmological controversies. http://www.newscientist.com/article/mg19025481.300 -exploring-stephen-hawkings-flexiverse.html?full=true.

14. http://www.skeptic.com/past-lectures/the-grand-design/. Hawking is fully aware of the role of subjectivity in our understanding of the world. In 2008, he explained that, in order to understand the universe's origins, it would be best to see the universe in terms of its most fundamental characteristic—how we experience it (referring to a classical or Newtonian experience of events). He argues that working backward from our perceptions would allow us to see the best possibilities for how a universe might have evolved. In essence, he is arguing that our perceptions of the world should be the starting point for any theory, despite him also arguing that his no-boundary theory solves the problem of "What came before the beginning?" http://www.new scientist.com/article/mg19826624.300.

6. Metacognition

1. http://www.nytimes.com/2010/02/11/science/earth/11climate.html.
2. Ibid.
3. http://www.huffingtonpost.com/2010/02/15/donald-trump-points-to-sn_n _462834.html.
4. P. Harrison, "Descartes on Animals," *Philosophical Quarterly* 42, no. 167 (April 1992), http://www.jstor.org/pss/2220217.
5. http://www.sagecenter.ucsb.edu/center.
6. M. Gazzaniga, *Humans* (New York: Ecco, 2008), 1–3.
7. M. Kukutani, "Memoir is Palin's Payback to McCain Campaign," *New York Times,* 2009, 11–14.
8. http://www.pbs.org/wnet/nature/episodes/bower-bird-blues/introduction /2109. For some wonderful photos of bowerbird "art objects," see http://www. flickr.com/photos/cuttlefishlove/galleries/72157623190971692/ and http:// ngm.nationalgeographic.com/2010/07/bowerbirds/morell-text.html.
9. P. Singer, *Animal Liberation* (New York: Avon Books, 1990), 10–12, 14–15, http://www.animal-rights-library.com/texts-m/singer03.htm.
10. A. Coghlan, "Animals Feel the Pain of Religious Slaughter," *New Scientist*, October 13, 2009, http://www.newscientist.com/article/dn17972-animals -feel-the-pain-of-religious-slaughter.html.
11. Ibid.
12. A good review of the central pros and cons in the Chinese room argument can be found at http://www.iep.utm.edu/chineser.
13. We assign credit to those who asked the correct questions and then relied on the computer to perform the complex calculations that led to an answer. We don't credit the computer with having come up with any of these ideas.

7. On the Level

1. Slime molds defy any particular category. They move around to feed themselves, like animals, but they also give rise to fruiting bodies that contain spores, like fungi. http://www.rice.edu/sallyport/2004/winter/features/low society/index.html.
2. S. Johnson, *Emergence* (New York: Scribner, 2002), http://www.worldcat.org /wcpa/servlet/DCARead?standardNo=068486875X&standardNoType=1 &excerpt=true.
3. T. Nakagaki, H. Yamada, and A. Tóth, "Intelligence: Maze-Solving by an Amoeboid Organism," *Nature* 407 (September 28, 2000): 470.
4. arxiv.org/abs/0912.3967.
5. You can argue that they're not strictly "unicellular" in the traditional sense, as swarms of slime mold (*Physarum polycephalum*) cells fuse together to form a plasmodium—a gigantic cell with multiple nuclei and no cell membranes between them.

6. http://www.nytimes.com/2010/01/26/science/26obmold.html?emc=eta1
&pagewanted=print>. A. Tero, S. Takagi, T. Saigusa, K. Ito, D. Bebber,
M. Fricker, K. Yumiki, R. Kobayashi, and T. Nakagaki, "Rules for Biologically Inspired Adaptive Network Design," *Science* January 22, 2010, vol. 327,
no. 5964, 439–42.

7. http://www.sciencedaily.com/releases/2010/05/100525202301.htm from May
26, 2010, in *Proceedings of the Royal Society B.*

8. http://en.wikinews.org/wiki/Brain_chemical_Serotonin_behind_locusts
%E2%80%99_swarming_instinct.

9. J. Dyer, C. Ioannou, L. Morrell, D. Croft, I. Couzin, D. Waters, and J. Krause,
"Consensus Decision-Making in Human Crowds," *Animal Behaviour* 75
(2008): 461–70.

10. http://www.nytimes.com/2007/11/13/science/13traff.html?pagewanted
=3&_r=1&ei=5089&en=0693ae5413eb2a67&ex=1352696400&partner=rs
syahoo&emc=rss.

11. Third Joint Action Meeting, July 27–29, 2009, Het Trippenhuis Tinbergenzaal, Amsterdam, Netherlands, http://web.mac.com/gknoblich/page4/assets
/JAM3_program.pdf; A. Roepstorff, C. Frith, and U. Frith, "How Our
Brains Build Social Worlds," *New Scientist,* December 2, 2009, http://www
.newscientist.com/article/mg20427370.500-how-our-brains-build-social
-worlds.html?full=true&print=true.

12. R. Selich, "The Feel of Joint Action," Third Joint Action Meeting, July 27–29,
2009, Het Trippenhuis Tinbergenzaal, Amsterdam, Netherlands, reference 13, p. 63, http://web.mac.com/gknoblich/page4/assets/JAM3_program
.pdf.

13. S. Ng, S. Han, L. Mao, and J. Lai, "Dynamic Bicultural Brains: fMRI
Study of Their Flexible Neural Representation of Self and Significant Others in Response to Cultural Primes," *Asian Journal of Social Psychology* 13
(2010): 83–91.

14. J. Heinrich, S. Heine and A. Norenzayan, "The Weirdest People in the
World?," *Behavioral and Brain Sciences* 33 (2010): 61–135, http://papers.ssrn.com
/sol3/papers.cfm?abstract_id=1601785.

15. A. McDowell, "Westerners vs. the World: We Are the Weird Ones," *National
Post,* August 21, 2010.

16. http://www.edge.org/3rd_culture/boroditsky09/boroditsky09_index.html.

17. http://www.amazon.com/Decoding-Universe-Information-Explaining-Ev-
erything/dp/067003441X. Whatever your position on such lofty issues, from
the ground-level perspective, I doubt that there are many, if any, people who
believe information simply doesn't exist when not presently detectable in the
physical domain, i.e., as bits of data or packets of neurotransmitters.

18. Z. Merali, "Spooky Steps to a Quantum Network," *New Scientist,* October 7,
2006, http://www.newscientist.com/article/mg19225723.600-spooky-steps
-to-a-quantum-network.html?full=true&print=true.

19. D. Robson, "Brain 'Entanglement' Could Explain Memories," *New Scientist,* January 12, 2010, 42, http://www.newscientist.com/article/dn18371
-brain-entanglement-could-explain-memories.html?full=true&print=true.

20. C. Agulhon, J. Petravicz, A. McMullen, E. Sweger, S. Minton, S. Taves, K. Casper, T. Fiacco and K. McCarthy, "What Is the Role of Astrocyte Calcium in Neurophysiology?" *Neuron* 59, no. 6 (September 25, 2008): 932–46.

21. A. Koob, "The Root of Thought: What Do Glial Cells Do?" *Scientific American*, October 27, 2009, http://www.scientificamerican.com/article.cfm ?id=the-root-of-thought-what.

22. http://faculty.washington.edu/chudler/facts.html.

23. F. Azevedo, L. Carvalho, L. Grinberg, et. al., "Equal Numbers of Neuronal and Nonneuronal Cells Make the Human Brain an Isometrically Scaled-Up Primate Brain," *Journal of Comparative Neurology* 513, no. 5, (April 10, 2009): 532–41, http://onlinelibrary.wiley.com/doi/10.1002/cne.21974/abstract. Despite the widespread quotes that the human brain contains 100 billion neurons and ten times more glial cells, using newer techniques for cell counting, these authors found that there were equal numbers of neurons and glial cells. "The ratios between glial cells and neurons in the human brain structures are similar to those found in other primates, and their numbers of cells match those expected for a primate of human proportions. These findings challenge the common view that humans stand out from other primates in their brain composition." Note the tendency to see ourselves as unique popping up even in how we have traditionally assigned ourselves a different brain composition (more glia) than other primates.

24. http://faculty.washington.edu/chudler/facts.html.

25. K. Smith, "Neuroscience: Settling the Great Glia Debate," *Nature* 468 (2010): 160–62. http://www.nature.com/news/2010/101110/full/468160a.html.

26. Ibid.

27. Ibid. For example, the calcium waves within astrocytes travel much slower than the more rapid neuronal firing necessary for controlling muscle movements. Slower and longer-lasting modes of action are more typical of the chemically mediated endocrine system.

28. Ibid.

8. Talking in Tongues

1. J. Decety, K. Michalska, Y. Akitsuki, and B. Lahey, "Atypical Empathic Responses in Adolescents with Aggressive Conduct Disorder: A Functional MRI Investigation," *Biological Psychology* 80 (2009): 203–11.

2. http://www.edge.org/3rd_culture/ramachandran/ramachandran_p1.html.

3. http://www.yale.edu/caplab/Main/Publications_files/sdarticle%282%29.pdf D. Lyons, L. Santos, and F. Keil, "Reflections of Other Minds: How Primate Social Cognition Can Inform the Function of Mirror Neurons," *Current Opinion in Neurobiology* 16 (2006): 230–34. According to Yale psychology professor Derek Lyons, although mirror neurons would seem to be the ideal neural substrate for imitation, "the puzzling fact is that monkeys simply do not imitate. Indeed imitation appears to be a uniquely human ability." A. Gopnik, "Cells That Read Minds?," *Slate*, April 26, 2007, http://www.slate .com/articles/life/brains/2007/04/cells_that_read_minds.html.

4. http://www.livescience.com/220-scientists-read-minds.html.

5. A. Goldman, *Simulating Minds: The Philosophy, Psychology, and Neuroscience of Mindreading* (London: Oxford University Press, 2006). 113.

6. J. Swanson, "Mirror Neurons Also Respond to Language and Sound," seedmagazine.com, September 20, 2006, http://seedmagazine.com/content/print/mirror_neurons_also_respond_to_language_and_sound/.

7. PBS special on mirror neurons: *Nova*, January 25, 2005, http://www.pbs.org/wgbh/nova/body/mirror-neurons.html.

8. http://www.livescience.com/220-scientists-read-minds.html.

9. M. Iacoboni, "Imitation, Empathy, and Mirror Neurons," *Annual Review of Psychology* 60 (2009): 653–70, http://www.adineu.com.ar/IMITATION%20EMPATHY%20AND%20MIRROR%20NEURONS%20IACOBONI.

10. H. Thompson, "Empathetic Mirror Neurons Found in Humans at Last," *New Scientist*, April 16, 2010, http://www.newscientist.com/article/mg20627565.600-empathetic-mirror-neurons-found-in-humans-at-last.html.

11. http://www.edge.org/3rd_culture/ramachandran06/ramachandran06_index.html.

12. http://www.salon.com/news/environment/mind_reader/2009/02/26/bernie_madoff. Another line of evidence against bundling together understanding the intentions of others and the experience of empathy is the ability of computer software programs to predict human emotions by analyzing a person's facial movements. In assessing some emotions such as joy and sadness, computers are already outperforming humans. At the same time, the computers have no "experience" of these emotions. That software can recognize emotions despite the computer having no emotional capabilities strongly suggests that the reading of another's mind neither requires nor leads to a component of empathy. For more information on facial recognition by computers, see L. Grossman, "The Bare-Faced Truth," *New Scientist*, May 26, 2012, 19.

13. N. Danziger, I. Faillenot, and R. Peyron, "Can We Share a Pain We Never Felt? Neural Correlates of Empathy in Patients with Congenital Insensitivity to Pain," *Neuron* 61, no. 2 (January 29, 2009): 203–12.

14. http://www.reproductive-revolution.com/archive/empathy-genetics.html. For additional references and discussion, see my article on empathy at Salon.com, http://www.salon.com/news/environment/mind_reader/2009/02/26/bernie_madoff.

15. http://www.reuters.com/article/2011/05/05/us-science-evil-idUS-TRE7442Q620110505.

16. P. Hof and E. Van der Gucht, "The Structure of the Cerebral Cortex of the Humpback Whale, Megaptera novaeangliae (Cetacea, Mysticeti, Balaenopteridae)," *The Anatomical Record*, published online November 27, 2006, http://www.wiley.com/bw/journal.asp?ref=1932-8486&site=1.

17. "Mixing Modalities Is More Than Meets the Eye Alone. Focus on: 'Multisensory Versus Unisensory Integration: Contrasting Modes in the Superior Colliculus,'" *Journal of Neurophysiology*, 97, no. 5 (May 2007): 3165 (editorial to Alvarado et al. article 3193-3205 in same issue), http://jn.physiology.org/content/97/5/3165.full.

18. The UCLA researchers concluded that the finding of mirror neurons in the medial temporal lobe suggests that there are multiple neural systems for mirroring and that they include areas not normally considered part of the classical mirror neuron territory.

19. http://seedmagazine.com/contest/article/mirror_neurons_also_respond_to _language_and_sound/. *Seed* magazine, September 20, 2006.

20. http://www.theglobeandmail.com/technology/science/researchers-develop -camera-that-will-show-your-mind/article565631/.

21. PBS special on mirror neurons: *Nova*, January 25, 2005, http://www.pbs.org /wgbh/nova/body/mirror-neurons.html.

22. http://www.pbs.org/wnet/nature/mozu/html/intro.html.

9. Under the Big Top

1. http://www.salon.com/health/col/bob/1999/12/06/dr_bob/index.html.

2. http://ajp.psychiatryonline.org/cgi/content/full/161/8/1433.

3. http://www.nytimes.com/2011/06/29/nyregion/judge-denny-chin-recounts -his-thoughts-in-bernard-madoff-sentencing.html.

4. http://news.cnet.com/8301-11386_3-10131643-76.html.

5. http://news.cnet.com/8301-11386_3-10131643-76.html.

6. http://www.newscientist.com/article/dn18266-brain-scan-reveals-who-will -keep-their-promises.html.

7. S. Begley, "Mind Reading Is Now Possible," *Newsweek*, January 12, 2008, http://www.newsweek.com/2008/01/12/mind-reading-is-now-possible. html.

8. http://www.ninds.nih.gov/disorders/lesch_nyhan/lesch_nyhan.htm.

9. http://www.wired.co.uk/magazine/archive/2009/06/features/guilty?page=all.

10. Ibid.

11. Ibid.

10. Consciousness Unexplained

1. For a good overview of the diagnosis and prognosis of altered states of consciousness, see http://ldysinger.stjohnsem.edu/thm_580_bioethics/09_coma -et-al/05_min-con-st.htm.

2. Persistent vegetative state is a diagnosis; permanent vegetative state is a prognosis. http://www.lifeexpectancy.com/articles/ashwal1.pdf.

3. For a further discussion of the persistent vegetative state, see "The Vegetative State: Guidance on Diagnosis and Management," in *Report of a Working Party.* (London: Royal College of Physicians, 2003).

4. http://archneur.ama-assn.org/cgi/content/full/64/8/1098#REF -NBS70001-4.

5. In a 2011 study, Owen and colleagues instructed five vegetative-state patients

to move their right hand. Two with preserved auditory processing exhibited increased activity in the appropriate left premotor cortex even though there was no visible movement of the right hand. The authors concluded that this was a sign of movement preparation and/or planning, and "may represent residual voluntary processing in these two patients." The important take-away from this article is the ongoing inability of the research team to arrive at a confirmatory methodology to definitively determine whether or not the activation reflects a conscious act or subconscious brain processing. T. Bekinschtein, F. Manes, M. Villarreal, A. Owen, and V. Della-Maggiore, "Functional Imaging Reveals Movement Preparatory Activity in the Vegetative State," *Frontier of Human Neuroscience* 5 (2011): 5. http://www.ncbi.nlm.nih.gov /pmc/articles/PMC3031991/.

6. B. De Jong, M. Willemsen, and A. M. Paans, "Regional Cerebral Blood Flow Changes Related to Affective Speech Presentation in Persistent Vegetative State," *Clinical Neurology and Neurosurgery* 99 (1997): 213–16.

7. D. Menon, A. Owen, E. Williams, P. Minhas, C. Allen, S. Boniface, and J. Pickard, "Cortical Processing in Persistent Vegetative State," *Lancet* 1 (1998): 734–37.

8. N. Schiff, U. Ribary, Diana Moreno, B. Beattie, E. Kronberg, R. Blassberg, R. Giacoino, T. C. McCagg, J. Fins, R. Llinas, and F. Plum, "Residual Cerebral Activity and Behavioral Fragments Can Remain in the Persistently Vegetative Brain," *Brain* (2002): 125, 1210–34.

9. http://www.northeastcenter.com/the-mohonk-report-disorders-of-con sciousness-assessment-treatment-research-needs.pdf.

10. http://www.bmj.com/content/306/6892/1597.abstract.

11. The actual status of his level of consciousness most likely fell into the general category of MCS. He clinically showed intermittently reproducible but inconsistent response to some commands (i.e., "Move your leg"), but no consistent intentional or functional communication could be established.

12. M. Monti, A. Vanhaudenhuyse, M. Coleman, M. Boly, J. Pickard, L. Tshibanda, A. Owen, and S. Laureys, "Willful Modulation of Brain Activity in Disorders of Consciousness," *NEJM*, February 8, 2010.

13. For a tragic but exquisite description of a life from the point of view of someone with locked-in syndrome, read *The Diving Bell and the Butterfly*, dictated by Jean-Dominique Bauby to his secretary by the use of eyeblinks (one blink for yes, two for no).

14. http://archneur.ama-assn.org/cgi/content/full/61/9/1357#REF-NNR40003 -21#REF-NNR40003-21. E. Kobylarz and N. Schiff, "Functional Imaging of Severely Brain-Injured Patients," *Archives of Neurology* 61 (2004): 1357–60.

15. http://www.newyorker.com/reporting/2007/10/15/071015fa_fact_groopman #ixzz1SIz2Tn3A.

16. A Rooper, "Cogito Ergo Sum," *New England Journal of Medicine,* 2010, no. 362. February 18, 2010, 648.

17. M. Sara, F. Pistoia, G. Cernera, and S. Sacco, "Comments and Opinions," *Archives of Neurology,* 65, no. 3 (March 2008). 418,

18. A. Owen, M. Coleman, M. Boly, M. Davis, S. Laureys, and J. Pickard, "Detecting Awareness in the Vegetative State," *Science* 313 (2006): 1402.

19. D. McCabe and A. Castel, "Seeing Is Believing: The Effect of Brain Images on Judgments of Scientific Reasoning," *Cognition*, 107(1), April 2008, 345–52.   http://psy.psych.colostate.edu/mccabe/McCabe%20Castel%20Brain%20Images.pdf.

11. Anatomy of a Thought

1. http://www.newscientist.com/article/mg20727711.300-size-isnt-everything-the-big-brain-myth.html.

2. http://www.sciencenews.org/view/generic/id/46057/title/Book_Review_The_Root_of_Thought_Unlocking_Glia—The_Brain_Cell_That_Will_Help_Us_Sharpen_Our_Wits,_Heal_Injury,_and_Treat_Brain_Disease_by_Andrew_Koob.

3. http://www.nervenet.org/papers/NUMBER_REV_1988.html#1.

4. C. Wanjek, *Bad Medicine: Misconceptions and Misuses Revealed* (New York: Wiley, 2002); http://home.ix.netcom.com/~suzumi/badmedicine_ch2.pdf.

5. Ibid.

6. S. Zola-Morgan, "Localization of Brain Function: The Legacy of Franz Joseph Gall (1758–1828)," *Annual Review of Neuroscience* 18 (March 1995): 359–83.

7. http://vichist.blogspot.com/2008/11/its-all-in-your-head-phrenology-and.html.

8. http://www.dailymail.co.uk/sciencetech/article-1385931/Intelligent-people-thicker-insulation-brains-wires-according-scientists.html#ixzz1OQh1zevF.

9. Their graph showing the various degrees of genetic contribution to various areas of the brain can be seen at http://www.technologyreview.com/biomedicine/22333/page2/.

10. http://www.jneurosci.org/content/29/7/2212.full.pdf+html.

11. H. Chung, M. Chou, and C. Chen, "Principles and Limitations of Computational Algorithms and Clinical Diffusion Tensor MR Tractography," *American Journal of Neuroradiology* 32 (January 2011): 3–13, http://www.ajnr.org/cgi/reprint/32/1/3.

12. http://www.bri.ucla.edu/bri_weekly/news_090311.asp.

13. Ibid.

14. http://www.pnas.org/content/97/8/4398.long.

15. http://www.jneurosci.org/content/29/7/2212.full.pdf+html.

16. http://www.dailymail.co.uk/sciencetech/article-1385931/Intelligent-people-thicker-insulation-brains-wires-according-scientists.html#ixzz1OQh1zevF.

17. http://www.newscientist.com/article/dn20453-easily-distracted-people-may-have-too-much-brain.html.

18. A. Newberg, "Religious Experiences Shrink Part of the Brain, *Scientific American,* May 31, 2001. "Religious Experiences Shrink Part of the Brain," *Scientific American,* May 31, 2011, http://www.scientificamerican.com/article.cfm?id=religious-experiences-shrink-part-of-brain.

19. A. Owen, R. Hayward, H. Koenig, D. Steffens, and M. Payne, "Religious Factors and Hippocampal Atrophy in Late Life," *PLoS One*, March 30, 2011.

20. "Religious Experiences Shrink."

21. Ibid.

22. A. Newberg, *How God Changes Your Brain: Breakthrough Findings from a Leading Neuroscientist* (New York: Ballantine, 2009).

23. S. Healy, and C. Rowe, "A Critique of Comparative Studies of Brain Size," *Proceeding of the Royal Society B* 22 274, no. 1609 (February 2007): 453–64. http://rspb.royalsocietypublishing.org/content/274/1609/453.full.

24. A. Vance, "In Pursuit of a Mind Map, Slice by Slice," *New York Times*, December 27, 2010, http://www.nytimes.com/2010/12/28/science/28brain.html?_r=1&ref=science&pagewanted=print.

25. Ibid.

26. http://www.youtube.com/watch?v=HA7GwKXfJB0.

12. Moral Character—Fact or Fiction?

1. http://www.telegraph.co.uk/technology/google/8520033/Stephen-Hawking-tells-Google-philosophy-is-dead.html.

2. http://www.nybooks.com/articles/archives/2011/mar/10/how-we-know/?pagination=false.

3. In 1971, 1300 academic neuroscientists attended the annual Society of Neuroscience meeting; in 2009 there were more than 25,000 practicing neuroscientists in attendance. To stand out in this crowd requires being noticed. If the squeaky wheel gets the grease, the hyperbolic claim is more likely to get the grants, tenure, paid lectureships, TV and radio appearances and book contracts. It's not surprising that "12 pillars of wisdom" sounds more like neuro-marketing than hard science.

4. Does the abstract nature of the mind put it into a different category of investigation where no amount of data can overcome the intrinsic problems of reason? For starters, if the self is a virtual construct created by our brain, who is it that has or doesn't have free will? The alternative—assigning free will to neurons and connections—also makes no sense.

   Lines of reasoning for determining the presence or absence of free will must begin with the presumption that these assumptions are freely chosen. But the validity of your initial assumptions hinges upon your conclusion. If you conclude that we do not have free will, you will have negated your starting assumptions and canceled out the conclusion. If choice is a prerequisite for determining whether or not we have choice, we are left stranded at the starting blocks of any philosophical argument concerning free will.

   Second, it is often argued that free will is incompatible with a purely materialistic physical universe. If cause and effect stems from the interaction of physical properties, it is hard to explain how free will can interrupt the causal chain in order to exert its own independent effect. The logical end point of this line of reasoning is that all is determined "from the beginning" and that anything that we do with our minds will be fully determined by prior physical interactions of our brain. If this is true, it would also follow that philosophy is nothing more than the rote recitation of predetermined arguments (hardly a comforting thought to those who have spent their academic careers contemplating free will). But if we don't have free will to choose our thoughts, we have no choice in our lines of reasoning and the evidence that establish our belief in materialism. The circularity: complete determinism precludes free will, which in turn precludes our ability to confirm a deterministic point of view.

A further deterrent is our present-day lack of understanding of the physical nature of mental sensations. As we don't know how to study purely mental phenomena at a physical level, we are left looking in the wrong place. A full unraveling of the nature of the subatomic particles that constitute your lover's fingertips won't tell you the feeling of her touch. You don't look for wetness in the atomic structure of water. (I sometimes think that studying the mind through scientific methodology is like looking for sexy lingerie in the home improvement section of a department store. You may acquire plenty of tools, but you will not find sexy.) A particularly popular error inherent in trying to match different levels of biological phenomena is the argument that an immaterial entity such as a mind cannot affect the physical world (the basis of the philosophy of epiphenomenalism). One need only look at the striking brain changes that can be seen on fMRI with the placebo effect or how information is continuously reshaping our brain. If and when we know how mental states and information "exist" at the level of physical laws, this argument will simply disappear as a problem, and remain only as a historical reminder of the foolishness of not admitting our ignorance.

Perhaps one day we will have a working hypothesis as to how the mind operates. At a bare minimum this will require understanding a host of questions difficult to express, let alone address—from how sensations are experienced and thoughts are created and stored, to the relationship between the individual brain and cognitive activity present only at the group level. In this book I have tried to present some of the larger theoretical issues that won't go away with our present ways of thinking about the mind.

5. http://www.nytimes.com/2009/04/07/opinion/07Brooks.html.

6. http://www.wjh.harvard.edu/~mnkylab/publications/recent/Cima.pdf.

7. http://www.nytimes.com/2009/05/01/us/politics/01souter.text.html?sq=obama%20%22empathy%22&st=cse&adxnnl=1&scp=3&adxnnlx=1313002949-H3c0PbTTSJU8Lq9xGOWmcw.

8. http://www.forbes.com/2009/05/04/supreme-court-justice-opinions-columnists-epstein_print.html.

9. S. Harris, *The Moral Landscape: How Science Can Determine Human Values* (New York: Free Press, 2010), 1–2.

10. Ibid., 125.

11. For an intelligent exploration of how to approach the notion of a good life, read O. J. Flanagan, *The Bodhisattva's Brain: Buddhism Naturalized* (Cambridge, Mass.: Bradford Books, 2011).

12. http://www.philosophynow.org/issue81/Ethics_Made_Easy_Feel_Good _Do_The_Right_Thing.

13. K. Liljenquist, C. Zhong, and A. Galinsky, "The Smell of Virtue: Clean Scents Promote Reciprocity and Charity," *Psychological Science* 21 (2010): 311–14.

14. http://www.princeton.edu/~harman/Papers/Virtue.html.

15. http://ndpr.nd.edu/news/23601/?id=1344.

16. To get a sense of the possible mechanisms underlying conscious effort to remain "honest," see J. Greene and J. Paxton, "Patterns of Neural Activity Associated with Honest and Dishonest Moral Decisions," *PNAS* 106, no. 30 (July 28, 2009): 12506–11, http://www.pnas.org/content/106/30/12506.full. This article is a superb example of how to carefully outline a study's potential limitations, and should serve as a model for how neuroscience papers should frame behavioral conclusions. *PNAS* is the Proceedings of the National Academy of Sciences.

17. F. Nietzsche, *Assorted Opinions and Maxims* (1879), http://www.maxmore .com/selftrns.htm.

18. J. Radowitz, "Brain Scans May Identify Slackers," *Independent*, May 2, 2102.

19. http://www.popsci.com/science/article/2011-10/biased-brains-help-humans -always-look-bright-side-life.

20. The subgenual region of the anterior cingulate cortex.

21. http://www.weizmann-usa.org/news/weizmann/Brave-Brains-Neural -Mechanisms-of-Courage.

22. A. Owen, "Putting Your Intelligence to the Ultimate Test," *New Scientist*, November 1, 2010, http://www.newscientist.com/article/mg20827841.300 -putting-your-intelligence-to-the-ultimate-test.html.

23. http://www.bbc.co.uk/news/health-11620971.

24. http://news.bio-medicine.org/medicine-news-2/Fibromyalgia-pain-isnt-all -in-patients-heads—new–brain-study-finds-7876-1/.

25. http://www.med.umich.edu/opm/newspage/2007/hmfibro.htm.

26. K. Grogan, "Healthy Times for Pfizer," *PharmaTimes*, November 5, 2003, Pharmatimes.com.

27. http://www.salon.com/news/environment/mind_reader/2009/06/11/fibro myalgia.

28. D. Wegner, *The Illusion of Conscious Will* (Cambridge, Mass.: Bradford Books, 2002), 2.

29. To hear an excellent discussion of the varying ways of interpreting such studies, check out this video: http://www.youtube.com/watch?v=rqCAIzuXhkY.

30. C. Soon, M. Brass, H. Heinze, and J. Haynes, "Unconscious Determinants of Free Decisions in the Human Brain," *Nature Neuroscience* 11 (2008): 543– 45, http://www.nature.com/neuro/journal/v11/n5/abs/nn.2112.html.

31. http://www.newscientist.com/article/mg20627541.900-picking-our-brains -how-powerful-is-the-subconscious.html.

32. G. Heyman, *Addiction: A Disorder of Choice* (Cambridge, Mass.: Harvard University Press, 2010).

13. Tell Me a Story

1. http://www.msnbc.msn.com/id/34109227/#.

2. http://www.telegraph.co.uk/journalists/rebecca-smith/7274179/Coma-victim-was-not-really-communicating-say-doctors.html; http://triplehelixblog.com/2010/03/what-is-consciousness-%E2%80%9Cmedical-miracle%E2%80%9D-provides-more-questions-than-answers/.

3. http://bmjopen.bmj.com/content/early/2011/02/16/bmjopen-2010-000039.short. Laureys and colleagues acknowledged the results might be biased, since the study was based on a select group and might not apply to locked-in patients in general. Patients also answered the survey with a caregiver or family member present, which may have influenced answers.

4. J. Kruger, and D. Dunning, "Unskilled and Unaware of It," *Journal of Personality and Social Psychology* 77, no. 6 (1999): 1121–34.

5. William James once argued that every philosophic system sets out to conceal, first of all, the philosopher's own temperament: that prerational bundle of preferences that urges him to hop on whatever logic train seems to be already heading in his general direction. This creates, as James put it, "a certain insincerity in our philosophic discussions: the potentest of all our premises is never mentioned. . . . What the system pretends to be is a picture of the great universe of God. What it is—and oh so flagrantly!—is the revelation of how intensely odd the personal flavor of some fellow creature is." http://nymag.com/arts/books/features/60120/.

Index